Business Guides on the Go

"Business Guides on the Go" presents cutting-edge insights from practice on particular topics within the fields of business, management, and finance. Written by practitioners and experts in a concise and accessible form the series provides professionals with a general understanding and a first practical approach to latest developments in business strategy, leadership, operations, HR management, innovation and technology management, marketing or digitalization. Students of business administration or management will also benefit from these practical guides for their future occupation/careers.

These Guides suit the needs of today's fast reader.

Jonas Rashedi

The Data-driven Organization

Using Data for the Success of Your Company

 Springer

Jonas Rashedi
Waldbronn, Germany

ISSN 2731-4758 ISSN 2731-4766 (electronic)
Business Guides on the Go
ISBN 978-3-031-20603-0 ISBN 978-3-031-20604-7 (eBook)
https://doi.org/10.1007/978-3-031-20604-7

© The Editor(s) (if applicable) and The Author(s), under exclusive licence to Springer Nature Switzerland AG 2023
Translation from the German language edition: "Das datengetriebene Unternehmen" by Jonas Rashedi, © Der/die Herausgeber bzw. der/die Autor(en), exklusiv lizenziert durch Springer Fachmedien Wiesbaden GmbH, ein Teil von Springer Nature 2022. Published by Springer Fachmedien Wiesbaden. All Rights Reserved.
This work is subject to copyright. All rights are solely and exclusively licensed by the Publisher, whether the whole or part of the material is concerned, specifically the rights of reprinting, reuse of illustrations, recitation, broadcasting, reproduction on microfilms or in any other physical way, and transmission or information storage and retrieval, electronic adaptation, computer software, or by similar or dissimilar methodology now known or hereafter developed.
The use of general descriptive names, registered names, trademarks, service marks, etc. in this publication does not imply, even in the absence of a specific statement, that such names are exempt from the relevant protective laws and regulations and therefore free for general use.
The publisher, the authors, and the editors are safe to assume that the advice and information in this book are believed to be true and accurate at the date of publication. Neither the publisher nor the authors or the editors give a warranty, expressed or implied, with respect to the material contained herein or for any errors or omissions that may have been made. The publisher remains neutral with regard to jurisdictional claims in published maps and institutional affiliations.

This Springer imprint is published by the registered company Springer Nature Switzerland AG.
The registered company address is: Gewerbestrasse 11, 6330 Cham, Switzerland

Foreword by Vanessa Stützle

Data is the currency of the digital age. The ability to handle data and use it effectively is essential in the twenty-first century. It ensures the future viability of organizations.

DOUGLAS, until a few years ago seemingly confined to the role of traditional retail icon, recognized the opportunities of the digital transformation early on and successfully expanded its online activities. Alongside the original store business, e-commerce has long since formed the second pillar of the company and is showing extraordinarily strong growth momentum. The digitization of DOUGLAS is succeeding because we collect and evaluate numerous data points and use them specifically for the further development of the business. A key success factor here is the BeautyCard. With over 47 million cardholders, it is one of the largest customer loyalty programs in Europe. By collecting this data on a large scale and analyzing it with the help of AI, we personalize our customer approach, offer a unique shopping experience, and inspire people about beauty.

Good technological solutions are the indispensable prerequisite for a data-driven organization. At the same time, tried-and-tested thought patterns and orientations suddenly lose their validity. Whether an organization successfully manages such a fundamental transformation depends not only on the technology, but also essentially on the organization's ability to change and learn. This requires openness to change, which also means being allowed to make mistakes. Digital transformation is only possible in such an open-ended context. It is central that organizations recognize the necessity of this change and mobilize their own potential. This also means rethinking the selection of the right talent, the organizational structure, and the management style. Feedback and openness to criticism are the be-all and end-all for leaders in the digital age.

Organizations that learn to deal with data are better equipped for the future and enable the full exploitation of their potential. Jonas Rashedi has accompanied the transformation of DOUGLAS over the last few years and very successfully implemented the data-driven infrastructure required for this.

In his book, Jonas Rashedi provides a clear guide to the process of becoming a data-driven organization. He shows the different facets that need to be considered on this path. In doing so, he makes an extremely valuable contribution to the question of how organizations can position themselves for the future in the age of digitization.

Chief Digital Officer of the Douglas GmbH   Vanessa Stützle
Düsseldorf, Germany

Preface

In our private lives, data and the insights it can provide are playing an increasingly important role. As a triathlete, I started out using simple stopwatches, and later added fitness trackers that helped optimize training with just a few pieces of data, such as heart rate. In the meantime, however, I use a wearable in conjunction with an app that collects and evaluates my physiological data around the clock, helping me to understand my body as well and in as much detail as possible. This enables me to adjust my sleep and time my workouts optimally. I see a very similar development with companies—here, too, there is enough data available about their own business, customers, competitors, and many other topics to make better decisions.

Where is the journey heading? The development into a data-driven organization is not merely an option for most companies, but a compelling necessity. However, the path to becoming a data-driven organization is not an easy one: Not only is an appropriate IT infrastructure required, but also suitable framework conditions in the company, such as a company-wide data strategy, suitable processes, and capable employees. Current studies also come to the conclusion that soft factors, such as corporate culture or the mindset of individual employees, are at least as relevant as the "hard factors" mentioned above.[1] All these factors must be planned and coordinated with each other.

Why is there no way around a data-driven organization? Ubiquitous facts, often packaged in buzzwords, such as increasing complexity and dynamics, higher volatility, and increasing unpredictability of developments present companies with massive challenges. In concrete terms, these developments manifest themselves, for example, in shorter product life cycles, which force companies to shorten and improve their own decision-making, development, and implementation work. In addition, much more data and the technologies needed to analyze it are available today—usually at manageable cost. This means that if you do not make your company a data-driven pioneer in your industry, another company will take over this role. Maybe even a player from outside the industry that takes advantage of the slowness of companies in the industry.

What exactly does it mean to be a data-driven company? In a data-driven company, decisions are supported by comprehensive and intelligent analyses of extensive amounts of data. This results in higher decision quality and improves the company's competitiveness. In other words, the company replaces "gut" decision-making or decision-making based on past data with decision-making based on current data. The insights gained from the data can be used, among other things, to reduce the company's own costs, improve value creation processes, and develop new products and services—and even innovative business models.

What does my company gain from being data-driven? In a nutshell: A better financial bottom line. A recent study by Capgemini concludes that companies that lead their industry in data usage generate about 22% more revenue than their competitors. Revenue per employee is even 70% higher.[2]

Does every company have to be data-driven? Every company can be data-driven. However, the question is to what extent the company wants

---

[1] Cf. New Vantage Partners (2019). Big Data and AI Executive Survey 2019. https://www.tcs.com/content/dam/tcs-bts/pdf/insights/Big-Data-Executive-Survey-2019-Findings-Updated-010219-1.pdf, p. 7. Accessed 12 Nov. 2021.

[2] Cf. Capgemini (2020). The data-powered enterprise. https://www.capgemini.com/no-no/wp-content/uploads/sites/28/2020/11/Data-powered-enterprise-report.pdf, p. 3. Accessed: 12.11.2021.

or has to work data-driven. As a triathlete, I can also make the following comparison: Anyone can complete a triathlon—the only questions are: Do I choose a small, a medium, or a large triathlon and how long will it take me to reach the finish?

How does this book help you on your way to becoming a data-driven company? My book supports you on the path to becoming data-driven. I don't claim that this is an easy path. You won't find recommendations on how to become a data-driven company right away in this book. Rather, I support you through a process based on my practical experience that breaks down the path to becoming a data-driven company into individual topics, but never forgets to take a holistic view.

How is this book structured? This book is divided into five chapters plus a conclusion. Chapter 1 highlights the key prerequisites and drivers of a data-driven organization. These are, for example, technological developments, changing framework conditions with regards to decision-making, and the market environment in front of the background of changing business models and customer needs.

Chapter 2 discusses the definition, characteristics, and functionality of a data-driven organization. There, I explain what data actually is, how a data-driven company functions, and what characteristics distinguish it.

An examination of the challenges and barriers to implementing a data-driven organization forms the subject of Chap. 3. To this end, I present a number of recent empirical studies and supplement the results with my own experiences.

Chapter 4 illustrates an easy yet fundamental 5-step process that gives an overview on how to manage data within an organization.

Chapter 5 is where I design a process model for implementing the data-driven organization and show you the fields of action that must be addressed and worked on in order to implement the data-driven organization.

You will notice while reading that this book has both theory and practice parts. However, I am limiting the theory to what I believe are the key points in order to establish a common understanding. Chapters 1 and 3 are theory-heavy, as they highlight the background for the data-driven organization and the challenges of implementation. Chapter 2 is also a foundation chapter, but intertwines theory and my practical experience. Chapter 4 presents the condensed findings from my years of consulting work and consequently contains theory only in a few places. My own experiences are enriched in these chapters with expert knowledge from my podcasts with various interlocutors.

In designing this book, I made a conscious decision to be brief. For this reason, I could not and did not want to illuminate every single aspect of the data-driven organization in detail. The book is more to be understood as a framework that needs to be further filled in at one point or another. My podcast "My Data is better than yours" is a good example of this.

And now I wish you much success in reading and implementing.

Waldbronn, Germany   Jonas Rashedi

Contents

1 Background and Drivers of the Data-Driven Organization   1
    1.1   Business Intelligence Development   1
    1.2   Drivers of the Data-Driven Organization   5
        1.2.1   Change in the Technological Environment   5
        1.2.2   Changed Decision Situation   7
        1.2.3   Changing Competition and New Business Models   10
        1.2.4   Changing Customer Behavior   12
        1.2.5   Drivers Summary   13
    References   15

2 Characteristics of the Data-Driven Organization   17
    2.1   Derivation of the Data-Driven Organization   17
        2.1.1   What Is Data?   17
        2.1.2   What Is a Data-Driven Business?   19
    2.2   What Do "Better" Choices Mean?   20
    2.3   Maturity Levels of Data-Driven Companies   22
    2.4   Properties of Data for the Data-Driven Organization   24
    2.5   Types of Analyses   26
    2.6   Advantages of a Data-Driven Company   27
    References   31

3 Challenges and Barriers of the Data-Driven Organization — 33
- 3.1 Empirical Studies on Challenges and Barriers — 33
- 3.2 Summary of Findings and Evaluation — 35
- References — 37

4 Process Model for Data Management — 39
- 4.1 The Five Steps — 39
- 4.2 Collect—Collect Data — 40
  - 4.2.1 What Is Data? — 41
  - 4.2.2 How Can We Differentiate Data? — 43
  - 4.2.3 Which Data from Which Sources Can Be Used? — 44
  - 4.2.4 More Data, More Knowledge? — 45
  - 4.2.5 How Do Data Silos Arise and How Do We Deal with Them? — 46
  - 4.2.6 What Criteria Are Relevant in the Choice of Technology? — 48
  - 4.2.7 What General Conditions Do We Have to Consider? — 49
  - 4.2.8 Guiding Questions for Collect — 50
- 4.3 Understand—Understanding the Collected Data — 51
  - 4.3.1 Why Is Understanding Central? — 52
  - 4.3.2 What Conditions Do We Need to Be Able to Understand? — 53
  - 4.3.3 What Must a Technical Preparation Look Like? — 54
  - 4.3.4 How Can We Tap into Data? — 55
  - 4.3.5 What Does Emotionalizing Data Mean? — 56
  - 4.3.6 How Can We Facilitate an Understanding? — 56
  - 4.3.7 Guiding Questions for Understand — 58
- 4.4 Decide—Decide on the Basis of the Collected Data — 58
  - 4.4.1 What Distinguishes a Data-Driven Decision from a Gut Decision? — 59
  - 4.4.2 What Types of Decisions Are Made in Companies? — 60
  - 4.4.3 What Are the Requirements for Making a Good Decision? — 61

		4.4.4	What Role Does the Time Factor Play in Decisions?	62

- 4.4.4 What Role Does the Time Factor Play in Decisions? — 62
- 4.4.5 How Can We Visualize Data? — 63
- 4.4.6 Data Versus Gut—Or Better in Combination? — 65
- 4.4.7 Guiding Questions for Decide — 66
- 4.5 Automate—Automation — 66
  - 4.5.1 Why Can't We Get Around Automation? — 66
  - 4.5.2 What Are the Technical Requirements for Automation? — 67
  - 4.5.3 What Added Value Does AI Create in the Context of Automation? — 68
  - 4.5.4 Is Automation Even More Than AI? — 69
  - 4.5.5 How Do We Manage to Transfer Our Findings into Processes in an Automated Way? — 70
  - 4.5.6 What Can Be the Causes of Resistance to the Data-Driven Organization? — 71
  - 4.5.7 Guiding Questions for Automate — 72
- 4.6 Summary — 72
- References — 72

5 Process Model for Implementing the Data-Driven Organization — 75
- 5.1 Overview — 75
- 5.2 Status Quo, Goals, and Data Strategy — 77
  - 5.2.1 Internal and External Analysis — 77
  - 5.2.2 Data Targets — 83
  - 5.2.3 Data Strategy — 86
  - 5.2.4 Stakeholder Integration — 91
- 5.3 Organization Model — 91
- 5.4 Process Model — 94
- 5.5 Example Projects — 95
  - 5.5.1 Self-Services and Real-Time Services at the Schwarz Group — 95
  - 5.5.2 Data & Analytics in B2B — 98

		5.5.3	Configuration of the Collaboration Between Data & Analytics and the Business Departments at a Fashion Company	99
		5.5.4	Re-launching Data & Analytics at a Content Provider in the Sports Sector	100
	5.6	Tools		102
	5.7	Data Culture		103
	5.8	Talent Management and Talent Strategy		107
	5.9	Data Governance		111
	References			115
6	**Closing Words**			117

1

Background and Drivers of the Data-Driven Organization

Abstract This chapter highlights the importance of BI in the context of the data-driven organization and identifies its key external and internal drivers.

1.1 Business Intelligence Development

The first use of the term Business Intelligence (BI) is attributed to Richard Millar Devens. He used it to describe the behavior of a banker who collected information about the competition in order to gain financial advantages through the knowledge gained. In the mid-twentieth century, a computer scientist from the IBM company used the term BI in a technical article. According to this article, the term refers to the potential for gaining insights resulting from technological possibilities (cf. Foote, 2017).

Figure 1.1 shows the origins and development of BI. The origins lie in so-called Management Information Systems (MIS). These were computer-based information systems that could be used to both collect and process information. Based on this information, analyses could be carried out, which helped the company management to make decisions. These

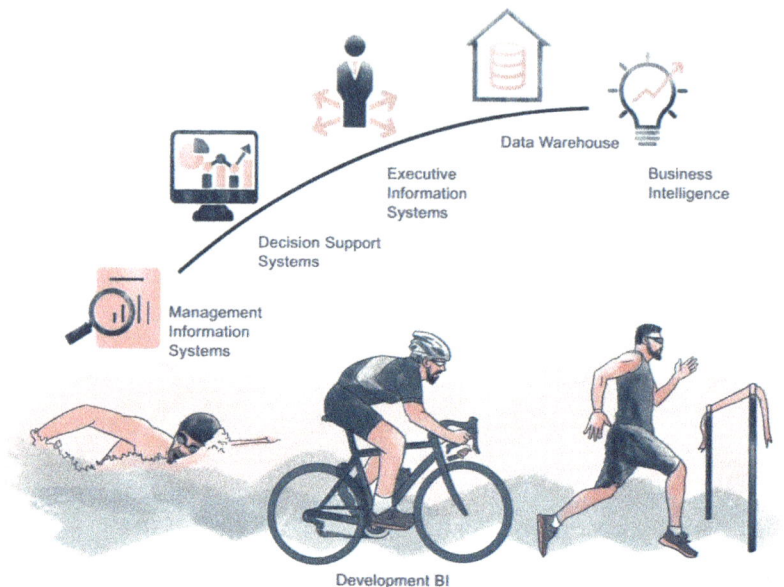

Fig. 1.1 Overview development BI (Source: Own representation)

systems came into use in the 1960s. At that time, there was also a vision of the "automatic decision generator."

Decision Support Systems, the next stage of development in the 1970s, provided support in the preparation and clear compilation of both operational and strategic information. These systems had static algorithms that could be used, for example, to implement "if-then" operations. Summations and average calculations, comparisons, or the creations of scenarios were also possible. Typical use cases were, for example, advertising budget or cash flow planning.

Executive Information Systems (EIS) were introduced in the late 1970s. These supported decision makers in areas such as marketing or customer management. Characteristic features of these systems were user-friendly interfaces with graphical displays for presenting the relevant information. Data warehouses were developed in the 1990s. These systems integrated data from various sources and answered interactive queries. Data warehouses also made a decisive contribution to the development and use of Big Data (cf. Reinkemeyer, 2018, p. 45; Foote, 2017).

1 Background and Drivers of the Data-Driven Organization

In the late 1990s and early 2000s, the term BI became popular. BI systems from numerous vendors were capable of generating reports and visualizing them in a suitable and appealing way. However, the operation of these tools required extensive know-how, so the tools were mainly used by experts. Subsequently, manufacturers modified the tools so that users without a technological background could also use them (BI 1.0). New technological possibilities marked the transition to BI 2.0 at the beginning of the twenty-first century. These technologies enabled the use of cloud-based solutions, the analysis of data in real time, and self-service systems through which decision makers could access information independently (cf. Heinze, 2020).

In today's world, there are different understandings of the term BI. The term "intelligence" is to be understood as the acquisition of information and knowledge as well as the goal-oriented collection and evaluation of data (see Grünwald & Taubner, 2009, p. 398). To date, no uniform understanding of the term has been established. For example, BI is understood to mean both the processing of data and information for corporate management as well as early warning systems and systems for storing information and knowledge. Another understanding of BI conceives of it as a process or filter for mastering the constantly growing volume of data (cf. Kemper et al., 2006, p. 3). If one attempts to systematize the multitude of different definitions, four different understandings can be distinguished from one another (cf. Kemper et al., 2006, pp. 3–4; Gluchowski et al., 2008, p. 91):

1. **The narrow understanding of BI**: The narrow understanding subsumes under the term BI only a number of core applications that support the decision maker. Core applications include, for example, Online Analytical Processing (OLAP), Management Information System (MIS) and Executive Information Systems (EIS).
2. **The analysis-oriented understanding of** BI: This understanding assigns not only core applications to BI, but all applications in which a person (e.g., the decision maker himself or a person preparing the decision) works directly with a system and has access to interactive

functions via an interface. Examples of such systems are text mining,[1] data mining[2] and ad hoc reporting.[3]

3. **The broad understanding of BI**: The broad understanding refers to all applications that are used directly or indirectly to support decision-making. This also includes, for example, applications for data preparation and storage as well as for the presentation of the insights gained. Examples at this point are Snowflake, Fivetran, Exasol, or Talend (software tools for bundling, analyzing, and sharing data via a central cloud platform), Fivetran (=cloud-based ETL[4] software).

4. **The process-oriented understanding**: While the three understandings mentioned above refer exclusively to tools and applications, this understanding sees BI as a process that uses data from various sources to tap potential and perspectives. This means that the company adapts to changing environmental situations through the use of techniques and tools and can identify both opportunities and risks at an early stage.

What is the significance of BI in the context of the data-driven organization? BI provides the basis for data-driven decisions. However, I have noticed that many companies only use BI to create financial reports. But these are not very data-driven, if at all. You could say they are more of a communication tool for certain decisions than a tool to help a company make better decisions—but that is what we are aiming for with the data-driven organization.

In concrete terms, this now means: Review your BI landscape and try to understand how it is currently set up, how it works and how it is used. Is it already helping to make data-informed decisions? If not, what steps

---

[1] Text mining means the analysis of documents with the aim of extracting information from these documents. Text mining is used to structure the unstructured information contained in documents and present it in a form appropriate for the user.

[2] Data mining is an umbrella term for a range of methods aimed at gaining insights from data. At its core, data mining involves the use of statistical and mathematical methods to identify relationships, patterns, trends, etc. in large volumes of data.

[3] Ad hoc reporting includes reports that are generated on an ad hoc basis rather than at a regular time.

[4] The abbreviation ETL stands for the terms Extract, Transform and Load and thus refers to operations that are performed on or with data. The objective of the process is to merge data from different sources and with different structures into a target database while adhering to a uniform format.

need to be taken to ensure that the BI landscape can contribute to making better decisions?

In some companies, the BI landscape is probably already getting on in years. A common problem here is that a "proliferation" has developed over time because, for example, change requests from individual users have been implemented or workarounds have arisen over time. This increases the complexity of the system, making it more difficult to operate and work with. This is especially true when there is a high turnover of personnel, as the original intentions or usage intentions have been lost. The result can be high time requirements for evaluations or the lack of important information. Older systems are also no longer adaptable to new requirements after a certain point in time and are consequently inflexible. As a result, the systems no longer meet the requirements placed on them, which not only reduces the acceptance of the system, but also, under certain circumstances, the acceptance of the generated results. In such a case, it may make sense to set up a completely new system—but to operate the old system in parallel until the new system has reached full performance.

1.2 Drivers of the Data-Driven Organization

1.2.1 Change in the Technological Environment

Crucial to the evolution of BI systems has been a series of developments in the technological field. Firstly, there has been an exponential increase in processor performance while production costs have remained constant. As early as 1965, Gordon Moore assumed that processor performance would double every year as part of the law named after him. Currently, there is no doubling every 12 months, but at least every 18 months (cf. Moore, 1965, pp. 114–117).

Secondly, falling costs for digital storage media are having a relevant influence on the development, since data storage is a mandatory condition for data collection, evaluation, and presentation. Here, too, there is an exponential development in terms of the available storage volume

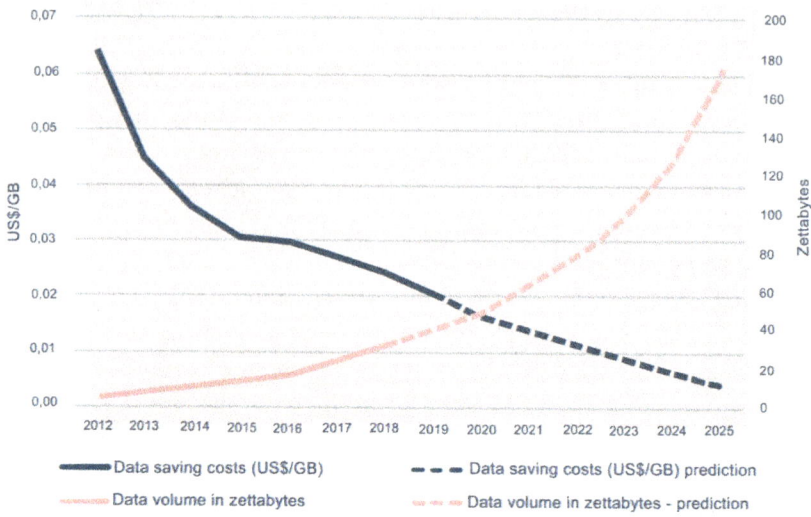

Fig. 1.2 Development of storage media costs (Source: Hammel & Wolf, 2020)

with simultaneously falling prices for storage. Furthermore, cloud solutions are gaining importance in the context of data storage. In 2019, for example, more data was stored in clouds than on local storage devices for the first time (see Hammel & Wolf, 2020, see Fig. 1.2).

The third trend is the development of the bandwidth that can be used for data transmission. As in the area of processor performance, there is also a regularity here (=Gilder's law), which assumes that the available bandwidth doubles every 6 months.

At the beginning of this chapter, I already mentioned the development of the amount of data available worldwide. This trend is also a driver for BI, since the more data available for analysis, the better the insights that can be generated. Causes for the increasing data volumes include the Internet of Things[5] and developments in connection with artificial intelligence.

In summary (see also Table 1.1), it can be stated that the technologies required for data processing in the context of BI have both increased in

---

[5] The Internet of Things refers to the networking of smart objects via the Internet. The objects can collect and evaluate data and perform activities based on the results of the analysis.

Table 1.1 Technological framework conditions

#	Area	Factors
1	Increasing performance at lower cost	Increasing processor performance Increasing bandwidth for data transmission Decreasing costs for data storage Data storage in the cloud
2	Increase in the amount of data available	Increasing volume of data worldwide More data for evaluation
3	Artificial intelligence	Evaluation of larger amount of data Gaining better knowledge

Source: Own representation

performance and fallen in price and have thus already become affordable for smaller companies. Furthermore, these technologies allow for more efficient and effective processes, as, for example, the use of cloud technologies reduces the likelihood of data silos arising and technologies such as artificial intelligence can be used to achieve better-quality results. For you, this means: You need both an overarching corporate strategy and people in the company who push technological trends on their own. To do this, these people need both the ability and the relevant knowledge, permission to act from the C-level, and sufficient qualified resources. I have also had good experiences with external partners, e.g., in implementing trends in my own projects. In doing so, it has proven to be expedient to choose rather smaller technology partners, as they are usually more flexible than large companies. Under certain circumstances, this puts you in a position to help determine the future product roadmap.

1.2.2 Changed Decision Situation

For companies and decision makers, the situation has changed in recent years. Firstly, the amount of data and information to be taken into account in decision-making has increased. The higher volume of data to be taken into account is due, on the one hand, to the fact that companies generally have more data at their disposal. On the other hand, decision-making situations have become more complex. "Complex" in this context means that more parameters have to be taken into account in

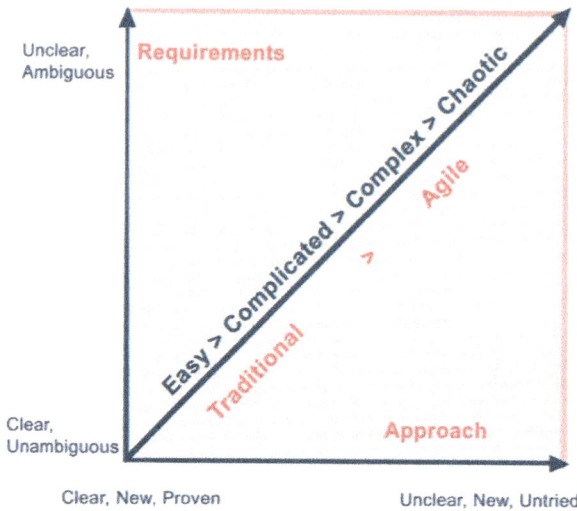

Fig. 1.3 Stacey matrix (Source: Angermaier, 2018)

decision-making (see also Fig. 1.3): As complexity increases, so does the number of factors to be considered for decision-making, as well as the relationships and interactions between these factors. Thus, simple decision-making situations are characterized by a high degree of transparency, as only a few factors can act and influence the situation. In complicated situations, the number of factors and the interrelationships between these factors increase. Typical examples of complicated tasks are optimization tasks. In these situations, however, traditional solution methods can still be used (e.g., operations research), but common sense or intuition is no longer sufficient to cope.

The more complex a situation becomes, the more interrelated factors have to be taken into account. Another characteristic feature of a complex situation is that influencing factors are continuously added or removed and, in addition, the nature and intensity of the relationships change. As a result, situations develop a momentum of their own. For the decision maker, such situations are no longer transparent; he can no longer recognize a connection between cause and effect. To solve complex problems, methods such as design thinking can be used as an agile approach—and data analytics can help uncover previously unrecognized

1 Background and Drivers of the Data-Driven Organization

relationships between factors (e.g., between a sociodemographic variable and the purchasing behavior of a target group).

In complex situations, clear cause-effect relationships can still be recognized, but this is no longer the case in chaotic systems. Proven solution approaches, such as setting up hypotheses and checking them, no longer work in chaotic situations. The only alternative action in such situations is (arbitrary) acting, observing, and perceiving as well as reacting.

> **Example: Complex Situation**
> Today, a manager faces more competitors with more products and a higher number of variants per product. In addition, he has responsibility for a larger task area. Overall, this means that several decision parameters are interlinked and influence each other, exposing the manager to a more complex decision-making situation.

Secondly, the time available for decision-making has decreased. This circumstance can be attributed, among other things, to ever shorter product life cycles, so that companies have to bring more new products to market per unit of time in order to remain competitive. This requires immediate reaction and adaptation to the constantly changing market environment.

The tighter time budget for decision-making is also reflected in the way managers made decisions just a few years ago and the possibilities that exist today in the digital age: For example, "back then" decisions were made on the basis of printed-out, meter-long reports. Today, the information required for decisions is available in real time in some cases, or decisions can even be made in real time by tools. The previous statements are summarized in Table 1.2.

Table 1.2 Framework conditions for decision-making

#	Area	Factors
1	More data to consider in decision-making	Companies have more data at their disposal. Decision-making situations are characterized by more parameters to be considered
2	Shorter time for decision-making	Shorter product life cycles

Source: Own representation

1.2.3 Changing Competition and New Business Models

The third area to be noted is that competition is also changing and new competitors are appearing with modified or new digital business models. The new business models use data and/or rely on massive use of technology. But first a brief digression: What is a business model?

A business model[6] describes the logic according to which a company functions. Specifically, a business model applies statements:

- The customers of a company (=customer segments)
- The benefits offered to the various customer segments (=value proposition)
- The way in which value is created in the company (=value creation architecture), and
- Of revenue generation and the costs incurred (=revenue mechanics)

We can now see that companies are using data and technology to change either one, several or all dimensions of their business model. Frequently, competitors from outside the industry in the form of technology companies are challenging established companies for customers and thus for sales.

The cause and trigger for this are developments in connection with Industry 4.0 and Marketing 4.0 as well as the platform economy: plants, machines, and material interact with each other in smart production environments via the Internet, coordinate themselves and exchange data (Industry 4.0). Some of the necessary production data comes directly from the customer, as the customer has either entered the data directly or provided it elsewhere (Marketing 4.0). It is easy to see that data is the essential basis for this.

---

[6] See, for example, the Business Model Canvas (BMC) by Osterwalder and Pigneur (2011).

1 Background and Drivers of the Data-Driven Organization

> **Example: New Business Models Through Industry 4.0**
> A customer of adidas can personalize almost all of the company's current shoe models, as well as other products, by choosing material and color and decorating them with embroidery. Production takes place in the so-called Speed Factory, the prototype of which the company put into operation back in 2016 and which allows for more efficient production. As a result, the individualized products can be offered with only a small surcharge (see Galer, 2018).

The developments described above affect all four dimensions of the business model. Airbnb, for example, is addressing a new customer segment of people who previously never considered staying in a hotel due to high costs. These customers are offered an unprecedented value proposition, as Airbnb has a very large selection of accommodation options with local flair. The value architecture is also changed, as a digital platform is used and Airbnb itself does not maintain any overnight accommodation capacity, but only acts as an intermediary. This also changes the revenue mechanics, as Airbnb generates its revenue through commissions. Overall, then, Airbnb is an overnight accommodation provider without its own beds—which represents a completely different logic and costs the traditional companies in the industry, i.e., hotels and guesthouses, revenue. In summary, there are two developments: the modification of business models by existing competitors and the market entry of new competitors (see Table 1.3).

Table 1.3 New competitors and new business models

#	Area	Factors
1	Existing competitors modify their business model	Adjustment of one or more dimensions of the business model
2	New competitors enter the market	Changed or completely new business models and value-creation processes

Source: Own representation

1.2.4 Changing Customer Behavior

The final influencing factor is a change in customer behavior. The changes in this area are extremely complex and can only be described in brief here (see Table 1.4). One important change is that more and more customers are currently prepared to disclose some of their data if they receive better services in return. Better services in this context could be, for example, services based on data or an individualized shopping experience. Another aspect is that more customers are shopping in online "measurable" channels. What is meant by this is that the behavior of customers shopping online is automatically recorded in the online channels and the resulting data is stored in digital form. This makes it easier for companies to process the data—as it does not first have to be (laboriously) digitized.

Customer behavior is also changing due to a higher degree of transparency: Customers can obtain a very good overview of the available range online, also with the help of comparison platforms. This is accompanied by a decline in customer loyalty. This is related to a change in the customer journey: In the offline sector, customers visited a retail store and then selected a product in this store that met their expectations or at least came close to them. At most, the customer usually visited one or two other stores. In online shopping, however, the customer journey changes: online, the customer does not first choose the store and then the product—but first the product and then the store, usually the cheapest. This leads to the customer choosing a different store in many cases for the next purchase.

Table 1.4 Changed customer behavior

#	Area	Factors
1	Lower loyalty	Changing buying habits (e.g., research offline, purchase online)
		Higher transparency of prices leads to purchase from the cheapest supplier
2	Customers want more channels	Customers learn from other industries and expect these amenities everywhere
3	Customers disclose data	Customers give data in exchange for services or better user experience

Source: Own representation

But the most important point, in my opinion, is that there are many more channels available for marketing these days. Customers are aware of these channels and also want to use all of them. For companies, therefore, the question is not: "Which of the channels do we use?" because all channels must be played on. However, an individual decision must be made for each customer as to which channel or which combination of channels should be used for a specific action. The data required for making this decision arises from customer contact: The company is therefore required to collect and evaluate data on channel usage in order to understand which customer can best be reached with which messages in which situations. For example, the same content can be delivered to customer A via e-mail, but to customer B at a different time as a push message via smartphone. At the same time, this 360-degree understanding also forms the basis for decisions by machines.

1.2.5 Drivers Summary

Table 1.5 brings together the influencing factors from the areas of technology, decision-making situation, competition and business models, and customers, and derives specific consequences for companies.

It is clear that these are general factors. There are other factors that apply to every industry and every company. Some factors may have a stronger impact, others may have little or no relevance. For example, the loyalty problem is very evident in online retailing due to greater transparency for the customer and a changed customer journey. In contrast, shortened decision cycles are not as relevant in this industry. In this respect, each company must decide for itself which of the factors mentioned should receive more in-depth attention when conducting its own analysis. The following questions can help you to do this:

- What are the key drivers in your industry for the areas listed?
- Are there more or other factors in your industry?
- What factors will influence your business and company tomorrow and the day after?
- What specific measures do these influencing factors require?

Table 1.5 Summary of drivers

Area	Concrete influencing factors	Consequence
Technology	Decreasing costs for hardware with simultaneously increasing performance	Even small businesses can afford IT infrastructure and become real competitors to large enterprises through data-driven behavior
		Empowering people in the company to drive issues forward
		Partnerships with (small) partners for the implementation of technology projects
	Increase in the amount of data available	More data allows the generation of more and better insights
	Artificial intelligence	Companies can not only evaluate larger amounts of data, but also gain better-quality insights
Decision situation	More data to consider in decision-making	Decision-making situations become more confusing and difficult
	Shorter time for decision-making	Decision makers need to consider more data and tend to have less time to do so
Competition and business models	Increasing use of data and technology in the context of Industry 4.0 and Marketing 4.0	Customer-centric business models and development of new target groups
		New value propositions for the customer
		Changed value creation as well as possibility of multi-sided platforms
		Changed revenue mechanics and opportunity for reduced costs
		Overall: Tougher competitive situation and risk of losing customers

(*continued*)

Table 1.5 (continued)

Area	Concrete influencing factors	Consequence
Customers	Customers shop in channels that can be measured online	More data about the customer is available
	Customers disclose more data	
	Greater transparency and a changed customer journey	Decreasing loyalty of customers
	Customers use different channels	Companies must be able to serve different channels
		Companies must determine the most effective and efficient channels for each customer

Source: Own representation

References

Angermaier, G. (2018). Stacey matrix, definition of terms. In Projektmagazin (Ed.), *Project management glossary*. Retrieved November 10, 2021, from https://www.projektmagazin.de/glossarterm/stacey-matrix

Foote, K. D. (2017). *A brief history of business intelligence*. DATAVERSITY. Retrieved November 10, 2021, from https://www.dataversity.net/brief-history-business-intelligence/

Galer, S. (2018). *How Adidas uses AI to manufacture personalized sneakers*. Production Online. Retrieved November 10, 2021, from https://www.produktion.de/digital_supply_chain/wie-adidas-mit-ki-personalisierte-sneakers-fertigt-316.html

Gluchowski, P., Gabriel, R., & Dittmar, C. (2008). *Management support systems and business intelligence: Computer-based information systems for professionals and managers* (2nd ed.). Springer.

Grünwald, M., & Taubner, D. (2009). Business intelligence. *Informatics Spectrum, 32*(5), 398–403. https://doi.org/10.1007/s00287-009-0374-1

Hammel, C., & Wolf, G. (2020). *New storage technologies or new business models? What drives innovation*. Technology Foundation Berlin. Retrieved November 10, 2021, from https://www.technologiestiftung-berlin.de/de/blog/neue-speichertechniken-oder-neue-geschaeftsmodelle-was-innovation-treibt

Heinze, J. (2020). *History of business intelligence.* Better Buys. Retrieved November 10, 2021, from https://www.betterbuys.com/bi/history-of-business-intelligence/

Kemper, H.-G., Mehanna, W., & Unger, C. (2006). *Business intelligence? Fundamentals and practical applications: An introduction to IT-based management support.* Vieweg.

Moore, G. E. (1965). Cramming more components onto integrated circuits. *Electronics, 38*(8), 114–117.

Osterwalder, A., & Pigneur, Y. (2011). *Business model generation: A handbook for visionaries, game changers, and challengers.* Campus.

Reinkemeyer, L. (2018). Digital transformation of internal business processes. *DIGITAL WORLD. The Business Magazine on Digitalization, 1*, 44–45.

2

Characteristics of the Data-Driven Organization

Abstract This chapter addresses the characteristic features of the data-driven organization, develops a definition of the term data-driven organization, and highlights the key benefits.

2.1 Derivation of the Data-Driven Organization

2.1.1 What Is Data?

Often the terms data, information, and knowledge are used synonymously. In this book, we differentiate the three terms as follows: Data is initially understood to mean only symbols and signs. Data are collected by different procedures, for example, by sensors of machines or by users. An essential characteristic of data is that it has no inherent meaning, but acquires it only through a context. Information, on the other hand, is knowledge about an object or a circumstance. Information is created by enriching data with context. Knowledge is created by combining different pieces of information that are available about an object or circumstance. In contrast to information, knowledge makes it possible to make an informed decision.

> **Added Value of Data at a Clothing Manufacturer**
> The evaluation of product reviews provides a very good basis for product development. From the product reviews, for example, information can be obtained on problems with product use, on product features that are perceived as good and those that are not, and on visual design, etc. Ultimately, this provides direct, unfiltered feedback from customers. Ultimately, this provides direct, unfiltered feedback from customers. Another example is support in positioning a product, as data can be used to identify what motivates customers to buy (cf. Rashedi & Feng, 2021).

Data can be differentiated with regard to a number of criteria. A first differentiation distinguishes between quantitative data (e.g., price of a product or size of a garment) and qualitative data (e.g., written product evaluation). Furthermore, a differentiation can be made between real data (e.g., address of a customer) and calculated data (e.g., contribution margin for customer A per year). From another point of view, a differentiation can be made between plain text data (e.g., Excel table with readable data) and non-plain text data (e.g., encrypted data).

Data is relevant for companies for the following reasons:

- Data forms the basis for analyses that help decision-makers make informed decisions.
- Data should be the basis for any kind of measure in the company. For example, address data or e-mail data are required for a direct marketing measure.
- Data allow the effectiveness of measures to be verified and are thus the basis of learning and better measures in the future.

2.1.2 What Is a Data-Driven Business?

A data-driven organization makes strong use of current data and intelligent analytics to inform decisions. A data-driven organization is characterized by the following attributes:

1. The importance of data for the company.
2. The type of data used.
3. Scope of data use.
4. The type of analyses used.
5. The ability to derive recommendations from analyses and actions from recommendations.

In my opinion, the most important point is the importance of data for the company: In a data-driven company, data is understood as a strategic asset that has a decisive influence on success and (long-term) competitiveness. In non-data-driven companies, on the other hand, data is understood as a tool in the operational area that can sometimes be used to support decision-making or to justify predefined statements. The second point relates to the type of data used: While in data-driven companies, real-time data or at least very current data is used, in other companies, past-related data is used to substantiate decisions. The extent of data use, the third point, also distinguishes data-driven and non-data-driven companies: While the former companies generally use data extensively for decision-making, the latter companies use it only on a case-by-case basis. A US management consultancy (import.io) describes the extent of data use in data-driven companies very aptly:

> Every employee should be expected to collect, analyze, and learn from data on a regular basis. Data should be shared and used for planning and reporting purposes along with internal monitoring against goals and objectives for telling your story. (White, 2015)

The fourth point is to be seen in connection with the analysis methods used. Here, the use of intelligent analyses using AI as well as other advanced analysis methods and standard tools from the BI toolbox, Excel evaluations and similarly simple analyses are contrasted. The fifth point relates to the company's ability to gain insights from the data and to use these insights profitably for the company (=value-based decisions).

> **Synergy Effects in a Data-Driven Company**
>
> Data can be the starting point for cross-functional collaboration: Ultimately, the functional areas of a company may work on different sub-goals, but these all pay into the overarching corporate goal. If it becomes transparent via data that the actions of two teams positively influence each other, this can be the starting point for cross-divisional collaboration in a project (cf. Rashedi & Feng, 2021).

It is also important that all five of the above points are considered for a data-driven company. For example, a company cannot be described as data-driven if it merely purchases the relevant software solutions and hires or trains data scientists itself, but does not view data as a strategic asset and develops and pursues a data strategy. In the case described, the data scientists would not even know exactly which use cases make sense for the company.

2.2 What Do "Better" Choices Mean?

I have already indicated that decision makers in data-driven organizations are empowered to make better decisions. But what do "better decisions" actually mean in detail? The consulting firm McKinsey looks at "better decisions" along four dimensions. The first improvement relates to the speed with which decisions can be made. This means that machines and algorithms can deliver analyses without delay, which is a particular advantage when decisions have to be made in real time. The second aspect is accuracy. For example, the consulting firm states that predictive models, provided they have been fed with the right data, can give very good indications of possible developments. This allows a data-driven company to deal effectively with the available resources. The aspect of accuracy is especially relevant in situations where small deviations cause strongly different results. The third aspect, which relates to reliability and consistency of results, is also important: While two different people may arrive at different results and decisions given the same data, the algorithm will always output the same results given the same input data. This increases the reliability of results and helps to increase confidence in the proposed decision or decision support. Fourth, the results produced by machines and algorithms exhibit a high degree of transparency because an algorithm always acts according to predefined rules (see McKinsey & Company, 2016, pp. 75–76). So if a machine makes a decision, it can be reproduced by a human being. The prerequisite for this is, of course, that the algorithm is known. In the case of a human decision maker, the decision made cannot necessarily be reproduced. AIs are currently an exception, at least to some extent. Their decisions are not always comprehensible, since a great deal of data is included in the model. In this case, the observer perceives the AI as a black box: a lot of data goes in, some operations are performed, and a result is presented at the end. However, with the GDPR, there are legal regulations for the transparency of AI and science as well as companies are working on the so-called explainable AI (="Explainable AI). Transparency is always relevant when several actors are affected by a decision. In this way, information asymmetries between the actors can be avoided.

In my view, McKinsey's four points need to be supplemented by two further aspects: First, by using AI, deeper insights into data can be gained and thus correlations and patterns can be uncovered that could never be gained by a human or standard analysis. Thus, not only are results produced more quickly and more data can be considered, but the results also exhibit higher quality. Second, decisions can be made not only on the basis of historical data. Rather, as I will explain in more detail in Sect. 2.5, concrete forecasts for future developments can be derived from this historical data.

Table 2.1 illustrates the total of six aspects of improved decision-making using the example of a marketing campaign

Table 2.1 Examples of better decisions

#	Improved feature of the decision	Example
1	Higher speed	A data-driven company can track the sales of a particular campaign in real time. Thresholds have been set in advance that the campaign must meet in order to be successful. If these values are not met after a few hours, the campaign can be canceled immediately and the company saves money for further implementation
2	Higher accuracy	The values obtained through the stated campaign can be determined down to the penny
3	Higher reliability	The data on the campaign is obtained directly via interfaces, without any media breaks. Transmission errors therefore do not occur
4	Higher transparency	Not only can the overall result be considered for the campaign, but it can be explicitly analyzed how the result came about: Who bought? How much was bought? What did the buyers do afterwards?
5	Higher quality	Via AI, correlations in campaign data can be identified that a human would not have discovered: e.g., that a certain group bought significantly more or less often
6	Forward-looking forecasts	For the next campaign, an intelligent analysis can already be used to determine probable sales. This allows different designs of the campaign to be compared with each other and then the best scenario can be chosen

Source: Based on McKinsey & Company, 2016, pp. 75–76

2.3 Maturity Levels of Data-Driven Companies

A number of different frameworks exist in the literature to describe the maturity of data-driven companies. In my opinion, one very good model differentiates between "data-driven" companies, "data-informed" companies and "data-inspired" companies (see Table 2.2).

Each company must identify for itself, as part of an analysis, which maturity level to strive for. In my career as a consultant, I have found that not every company necessarily has to achieve the highest maturity level (=data-driven). For example, it may be sufficient for a company to achieve

Table 2.2 Overview maturity levels

Designation	Data-inspired	Data-informed	Data-driven
Existing data	The company engages in trend spotting and uses various data sources to anticipate future customer behavior	All employees are informed about the current performance of the company and the product performance, so that the corporate strategy can be optimized	The company has exactly the data it needs to make decisions
Type of decision-making	Data from various sources are used to generate hypotheses Data are interpreted and enriched with considerations or opinions beyond their actual significance	Data is considered in the context of decision-making Data is sometimes viewed critically	Decisions are made solely on the basis of the data generated The data is trusted

Source: Kumar, 2020

the data-informed maturity level because, for example, not all data is or can be available (digitally), not every decision has to be made in a data-informed manner, or the company is not yet far enough along in the digital transformation to be able to be data-driven and corresponding projects cannot even be implemented yet. Other companies, such as those in the automotive industry in Germany, tend to be highly optimized and will strive to achieve the highest level of maturity.

But if the goal is not necessarily to realize the highest level of maturity, what can you use to guide you and your company? How do you know what represents the right maturity level for your company? This question must always be considered based on the company's individual value chain: Data & Analytics should always start at the points that add the most value to the business. The following example shows that a use case that makes sense for many companies does not necessarily make sense for your own company.

> **Company-Specific View of Data and Analytics**
> Use cases for data science must always be considered on a company-specific basis. For some companies, for example, predictive maintenance can generate added value. However, if a company has only very few high-quality machines with low downtimes, then predictive maintenance is not a sensible approach: Predictive maintenance generates statements about possible future failures on the basis of past data. However, if only a few data on failures are available, no valid statements can be made about future failures (cf. Rashedi & Wernicke, 2021).

The example also makes it clear that data science projects fail in most cases not because of a missing or unsuitable model, but because of an inadequate data basis. In this respect, it must be questioned when designing the business case whether sufficient data is available and whether data represents the right way to solve the problem or create added value for the company.

2.4 Properties of Data for the Data-Driven Organization

Data-driven organizations are characterized by their data having the following properties (see Andersson, 2016; Hebbar, 2019; Preimesberger, 2020, among others):

- **Accuracy**: The data to be used must be correct. Otherwise, the well-known principle from computer science "garbage in, garbage out" applies.
- **Timeliness**: The data used in the analyses must also be up-to-date. This applies in particular to future-oriented analyses, i.e., predictive and prescriptive analyses.
- **Relevance**: The data must be relevant. Ultimately, it is not important to have as much data as possible (Big Data), but to have the data that is meaningful for answering the current question.
- **Linkable**: The linkability aspect refers to the form of the data. Specifically, this means that the data must be available in a form that

enables it to be merged with other data. Possible solutions for this are, for example, relational databases or NoSQL stores.
- **Democratic**: The aspect of data democracy refers to the data culture in the company. Specifically, this means that no one in the company is "sitting on" their data, but that there is a willingness to share it with other people and areas. This is the prerequisite for being able to merge the data available in different areas in a company.
- **Processable**: Finally, it must be possible to query and process the existing data, i.e., the company must have suitable tools for filtering, grouping, and aggregating data. This is the prerequisite for extracting the data required to answer the question from extensive raw data.
- **Understanding**: All employees in the company should have the same understanding of the existing data. If this is ensured, it is possible to work optimally with the data. Otherwise, misunderstandings will arise if, for example, it is not clear whether the "sales" date refers to sales with or without sales tax.

The aspects listed are relevant because the competitiveness of companies in the future will depend to a large extent on the ability to make data-driven decisions. Companies must be able to collect, process, communicate, and activate the data relevant to the company. However, this requires up-to-date and accurate data, otherwise the trust of employees and also customers in the data will dwindle. Similarly, it is not very effective if a company has current and accurate data, but no one in the company knows about this data or no one has the ability to activate this data and make it usable for the company.

But even here, a company does not necessarily have to have maximum values for all individual criteria. The requirements for the proficiency can also differ depending on the area of the company: For example, online marketing tends to require data to be highly up-to-date and shareable, while accuracy is not as important. In the financial sector, however, accurate and shareable, but not necessarily up-to-date data is required in connection with monthly and annual financial statements.

2.5 Types of Analyses

With regard to the type of analysis, four variants can be differentiated. They can all be subsumed under the heading of "data analytics." Descriptive analytics is the first type. They involve the evaluation of historical, structured data. The statements answer the question "What happened in the past?". Descriptive methods are often used in connection with BI systems (Fig. 2.1).

The second variant, diagnostic analytics, also uses historical data for analysis. However, these methods are not only aimed at describing facts, but also at explaining the causes of developments. This can be achieved by identifying patterns. This makes it possible to justify the deviations identified in the course of descriptive analysis. Methods such as data mining, correlation analyses[1] or data discovery are used.[2]

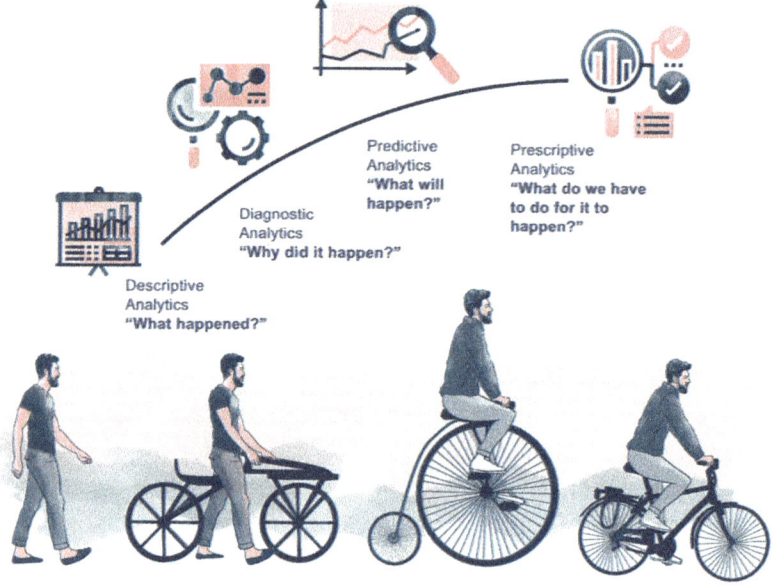

Fig. 2.1 Maturity model for data analytics (Source: own representation based on Amann et al., 2020, p. 253)

---

[1] Mathematical relationships between variables can be identified via a correlation analysis. However, no causal relationships can be proven via a correlation analysis.

[2] Data discovery involves searching for objects or patterns in data from different sources.

Table 2.3 Examples of analysis methods

Analysis type	Example from online retail
Descriptive analysis	Analysis of sales figures per product category and comparison of deviations to previous periods
Diagnostic analysis	Finding explanations why some products sold more and others sold less or not at all
Predictive analysis	Forecast which products will be most popular with customers in the future and therefore most in demand
Prescriptive analysis	Conduct scenario analysis to identify the most appropriate ways to use the advertising budget to "push" what are believed to be the most popular products

Source: Own representation

Predictive analytics go one step further: They can use the patterns identified in the data records to make statements about probable future behavior. Predictive methods use both mathematical and statistical methods, such as decision trees or regression analyses. Specifically, predictive methods identify possible future events and determine the probability of their occurrence. Both past data and current or real-time data are used.

Prescriptive analytics, as the last variant, is aimed not only at predicting future events but also at making statements on how the company can influence events to occur or not to occur. Methods such as simulations, scenario analyses, or if-then analyses using AI are used (see Boßow-Thies et al., 2020, pp. 11–13; Romeike & Hager, 2020, pp. 335–336). Table 2.3 shows examples of the analysis methods using the example of an online store.

2.6 Advantages of a Data-Driven Company

To be able to demonstrate the advantages of a data-driven company, we again use the business model approach already presented. A data-driven company can change all four dimensions of its business model (customer segments, value proposition, value architecture, and revenue mechanics).

Firstly, data-driven companies can address the customer dimension and tap into new customer segments. This is made possible by automated and thus more cost-effective processes, which enable services to be offered at lower prices and also less affluent customer segments to be tapped. One example is the company Rightmart, which specializes in social law and can offer services at very low prices, for example, through extensive use of technology. Furthermore, there is the possibility of better interaction with customers in the customer area, provided that data about the customer is generated and used in a targeted manner (data-driven marketing).

In terms of the value proposition, a data-driven company can offer its customers new services, focusing on the needs of the customer. Examples of such customer-centric business models are the so-called fintech companies in the banking industry. These do not offer the entire services of banks, but focus on one or a few aspects of the value chain of banks. Examples in Germany are Auxmoney (B2B lending), N26 (account management via smartphone) or Trade Republic (stock trading). New offerings for customers can also refer to data-driven benefits and services that are offered either as a supplement to a physical product or detached from it. At their core, these services are based on the company collecting and evaluating usage data and data on customer behavior and offering customers added value based on this. Another example is comparison platforms that offer customers transparency about services in different areas—and will very likely offer their own services in the future. Data-driven companies also have the ability to identify starting points for improvements and optimization. In a data-driven company, this can be done, for example, through sentiment analysis[3] in social media. Another example of added value to customers in connection with the value proposition is shown in the following example.

---

[3] The objective of a sentiment analysis is to analyze the emotional attitude of people towards an object such as a product or a company from texts.

> **Added Value for Customers with a Data-Driven Business Model**
> Typically, new vehicle customers want to configure their own vehicle and tailor the equipment to their personal needs. However, the consequence of an individual configuration is that such a vehicle cannot be delivered immediately, but sometimes very long delivery times arise. The customer is therefore faced with a trade-off between immediate availability of an "off-the-shelf" vehicle and availability of an individually configured vehicle only at a much later date.
> By evaluating both configuration data from car dealers and configurations made via the website, an automaker can identify which configurations are particularly popular with customers in a country. Vehicles can be pre-produced according to these configurations. During the sales pitch, the customer can be presented with a vehicle that already to the customer's needs can be offered for immediate purchase. In addition, the automaker can integrate higher-value equipment into the pre-produced vehicle configurations. The customer may pay the extra price for these, as the vehicle is immediately available in this configuration (cf. Rashedi & Wernicke, 2021).

A data-driven company can also address the value creation dimension of the business model. A central aspect in this context is the extensive use of technology. Examples of this are the FinTech companies already mentioned. But there are also corresponding companies in the retail sector, such as About You, which is breaking new ground in terms of marketing: The company relies on personalization, which is reflected, for example, in product assortments adapted to the user. Another aspect of value creation is the integration of partners, for example, via platforms. An example of this is the Airbnb platform, which connects not only providers and consumers of accommodation, but also, for example, providers of cleaning services or local scouts. At the same time, of course, this also changes the company's value proposition.

Furthermore, data-driven companies have the opportunity to implement optimizations. This can take place both in individual areas of the company (e.g., greater efficiency through data-driven marketing) and at the level of the company as a whole (e.g., control of the entire company). A data-driven company can continue to build a better picture of the customer and their needs. For example, in a data-driven organization—taking into account requirements related to data privacy—offline and online data can be merged and matched so that the company has the matching online profile to the analog shopper in the store. And last but not least, the management of a data-driven organization can make decisions with a higher quality. This is due to the fact that objective data is significantly more reliable than people's experiences, opinions, interpretations, or assessments. The decisions relate not only to the actual production of services, but ultimately to all processes taking place in the company.

In terms of revenue mechanics, it should be noted that companies have other ways of generating revenue. For example, prices for a service can be justified on the basis of actual use. One example in this context is the Rolls-Royce company, which no longer sells aircraft engines to customers but instead provides them in return for payment of a usage-based fee.

Figure 2.2 summarizes the opportunities for business model change.

In this context, we again have to ask ourselves the question of the (right) maturity level. Presumably, no company has to have the highest degree of maturity in all the areas listed. The maturity level of the individual company divisions or functional areas must always be considered in comparison to competitor companies from one's own industry or technology companies from outside the industry that are entering one's own

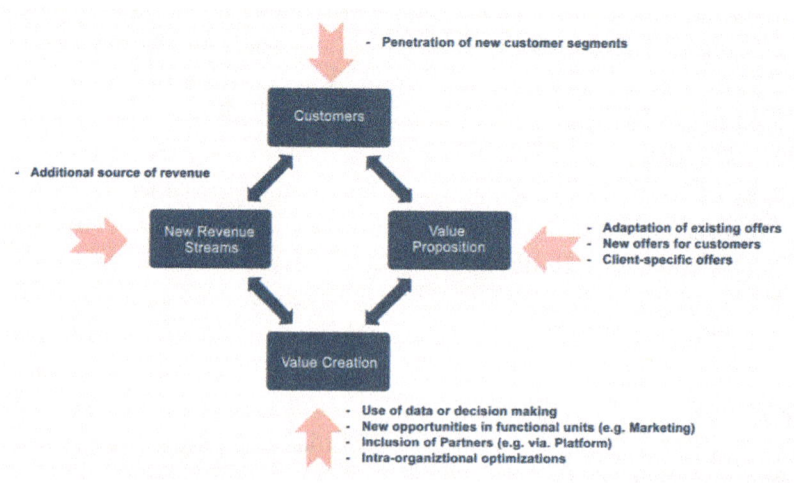

Fig. 2.2 Business model innovations in data-driven companies (Source: own representation)

industry for the first time: A company does not necessarily need an 800-horsepower engine in a vehicle, provided the competitors themselves do not have Formula 1 bolides, but only Formula 3 cars with far less horsepower.

References

Amann, K., Petzold, J., & Westerkamp, M. (2020). *Management and controlling: Instruments - organization - goals - digitalization*. Springer Gabler.

Andersson, R. (2016). *4 characteristics of data-driven organizations - and how to get started*. IBM Sverige - THINK Bloggen. Retrieved November 10, 2021, from https://www.ibm.com/blogs/think/se-sv/2016/04/15/4-characteristics-of-data-driven-organizations-and-how-to-get-started/

Boßow-Thies, S., Hofmann-Stölting, C., & Jochims, H. (2020). The oil of the 21st century - Strategic use of data in marketing. In S. Boßow-Thies, C. Hofmann-Stölting, & H. Jochims (Eds.), *Data-driven marketing* (pp. 3–26). Springer Gabler. https://doi.org/10.1007/978-3-658-29995-8_1

Hebbar, P. (2019). 10 indicators of a truly data-driven organization. *Analytics India Magazine*. Retrieved November, 2021, from https://analyticsindiamag.com/10-indicators-of-a-truly-data-driven-organisation/

Kumar, S. (2020). *Data-driven, data-informed & data-inspired - old ideas, new words*. Retrieved November 10, 2021, from https://towardsdatascience.com/data-driven-vs-data-informed-vs-data-inspired-740eaaec6263

McKinsey & Company. (2016). *The age of analytics: Competing in a data-driven world*. Retrieved November 10, 2021, from https://www.mckinsey.com/~/media/McKinsey/Industries/Public%20and%20Social%20Sector/Our%20Insights/The%20age%20of%20analytics%20Competing%20in%20a%20data%20driven%20world/MGI-The-Age-of-Analytics-Full-report.pdf

Preimesberger, C. (2020). *Nine key factors that identify truly data-driven organizations*. EWEEK. Retrieved November 10, 2021, from https://www.eweek.com/it-management/nine-key-factors-that-identify-truly-data-driven-organizations/

Rashedi, J., & Feng, T. (2021). *Adidas (Tiankai F.) - Why analysts can do much more than Excel and dashboards*. MY DATA IS BETTER THAN YOURS. Retrieved November 10, 2021, from https://mydata.podigee.io/47-tiankai-feng-analysten

Rashedi, J., & Wernicke, S. (2021). *Sebastian Wernicke & Jonas Rashedi - Data Science on the ground*. MY DATA IS BETTER THAN YOURS. Retrieved November 10, 2021, from https://mydata.podigee.io/38-data-science

Romeike, F., & Hager, P. (2020). *Success factor risk management 4.0: Methods, examples, checklists practical handbook for industry and commerce*. Springer Gabler. https://doi.org/10.1007/978-3-658-29446-5

White, D. (2015, May 20). *How (and why) to become a data-driven company*. Medium. Retrieved November 10, 2021, from https://medium.com/@david_white/how-and-why-to-become-a-data-driven-company-b758733e38ee

3

Challenges and Barriers of the Data-Driven Organization

Abstract This chapter examines challenges and barriers to implementing a data-driven organization. Three publications with an international focus are used for the analysis.

3.1 Empirical Studies on Challenges and Barriers

The first empirical study from 2011 aimed to find out how companies proceed in order to convert data and information into insights and concrete measures. The study was a collaboration between IBM and MIT Sloan Management Review. Insights were gained through expert interviews with business executives. Sloan Management Review identified a total of eleven barriers to implementation. The biggest six challenges are (cf. LaValle et al., 2010, p. 25):

1. Lack of understanding regarding the use of analytics to improve their own business.
2. Lack of "management bandwidth due to competing priorities."

3. Lack of internal skills.
4. Lack of data access.
5. Existing culture does not support data sharing.
6. Ownership of data or governance is unclear.

As can be seen from the list and is also explicitly stated by the authors of the study, the greatest challenges are not to be seen in connection with the availability or use of technologies. Rather, management-related and cultural aspects represent the greatest obstacles. For example, the publication states that it is a big and hard change for a manager to change his or her decision-making behavior, i.e., to no longer make decisions based on personal experience, but to make decisions based on data (see LaValle et al., 2010, p. 25).

The second study from 2017 was aimed at identifying patterns for building a data infrastructure at data-driven companies. In particular, it sought to identify best practices. The research used a mixed-methods approach consisting of a written survey of business intelligence and data professionals as well as telephone interviews with users, business sponsors, and BI/analytics experts. Participants in the survey belonged to various industries and companies of different sizes. Around two-thirds of the participants were from the USA, while the remaining respondents were predominantly from Europe, Canada, and Central and South America (cf. Halper & Stodder, 2017, p. 3).

The empirical study identifies a total of twelve barriers on the way to a data-driven organization. The biggest problem identified is a lack of support from business executives and corporate strategies. It is stated that the path to a data-driven organization would be easier to tread if business executives could be won over for support. These individuals could assist in formulating the vision and provide both necessary resources and other organizational support. Without support from business executives, the research found, more time would be needed. The second challenge, seen by as many as 37% of respondents, is accessing and integrating those data that the business needs to gain real insights. The second challenge can be based on different causes:

- Technological problem: The organization does not have the appropriate tools to integrate the data.
- Personnel problem: The organization does not have the appropriate personnel skills to integrate the data that exists in the data silos.
- Organizational problem: Insufficient financial resources or user refusal to share data.

Insufficient skills are seen as the third challenge. Skills are needed both in connection with data management and analysis and with regard to the implementation of visual analytics, i.e., the accomplishment of back-end processes (e.g., integration of analytics into applications) (cf. Halper & Stodder, 2017, pp. 9–10).

Also from 2017, is a study by New Vantage Partners, a consulting firm in the field of data-driven business innovation. Executives from 50 Fortune 1000 companies were surveyed for this study. The objective of the survey was to identify factors that significantly influence the acceptance, investment, and success of Big Data. It is not clear from the survey whether only organization-related or cultural barriers were explicitly queried or whether—similar to the survey described first—technological factors do not represent decisive barriers on the way to a data-driven organization. According to the study, the most important challenges are a lack of strategic orientation of the organization as a whole and too little understanding and acceptance of data at the level of middle management (New Vantage Partners, 2019).

3.2 Summary of Findings and Evaluation

Summarizing the results of the three studies, it can be noted that the majority of the identified challenges are in the areas of

- Strategy (e.g., lack of leadership support),
- Acceptance and understanding (e.g., lack of understanding regarding the use and added value of data), and
- The skills and resources available in the company

concerns. Although a great many challenges and barriers were also identified in the area of data itself, nevertheless only one concretely data-related problem (=difficulty in accessing data) can be found in the top three challenges of the studies. In concrete terms, this means that the challenges in connection with the realization of the data-driven organization do not lie in the area of data, but relate to leadership, resources and competencies, and corporate culture.

Based on my experience, I would like to add two concrete experiences in the DACH region to these findings. The first is silo thinking in companies. Data silos can arise for a variety of reasons. For example, if the individual departments in a company work with different tools and do not store the data centrally, but instead store it in their own databases. Another cause can be that data on the same customer is created at different touchpoints during the customer journey and this data is stored in different places and/or processed by different areas of the company. Regardless of how data silos are created, the consequence is that parts of the data available in the company cannot be accessed for evaluation. In many cases, the data silos are also likely to be the cause of the difficulty in accessing data cited in the studies.

Another point from my practical experience is the choice of the right technology. There are several reasons for this as well, for example, the high number of available solutions. For companies, the main question here is whether a complete solution with a large number of integrated solutions (all-in-one solution) should be purchased or whether the best solution available on the market should be purchased for the individual subtasks (best-of-breed approach). All-in-one solutions, which are usually offered by an established manufacturer, are generally associated with lower licensing costs and the interface problems between the individual applications are eliminated. However, these solutions do not necessarily meet the specific requirements of the company. With the best-of-breed approach, individual requirements can be taken into account, but the individual solutions have to be harmonized, which can result in high one-off costs—which are first amortized. Which must first be amortized. At the same time, further costs can also arise during operation, e.g., adaptation in the event of changing interfaces.

References

Halper, F., & Stodder, D. (2017). *What it takes to be data-driven: Technologies and best practices for becoming a smarter organization.* Retrieved November 10, 2021, from 7https://media.bitpipe.com/io_14x/io_141315/item_1674359/TDWI_BPReport_Q417.pdf

LaValle, S., Lesser, E., Shockley, R., Hopkins, M. S., & Kruschwitz, N. (2010). *Big data, analytics and the path from insights to value.* MIT Sloan Management Review. Retrieved November 10, 2021, from https://sloanreview.mit.edu/article/big-data-analytics-and-the-path-from-insights-to-value/

NewVantage Partners. (2019). *Big data and AI executive survey 2019.* Retrieved November 10, 2021, from https://www.tcs.com/content/dam/tcs-bts/pdf/insights/Big-Data-Executive-Survey-2019-Findings-Updated-010219-1.pdf

4

Process Model for Data Management

Abstract This chapter shows a simple yet holistic process model for data management that can be used in most organizations as a system to work more effectively with data.

4.1 The Five Steps

The process model for a data-driven organization can be mapped in five steps. "Collect" represents the first step and refers to the collection, storage and preparation of data for further processing. Here, the first step is to identify the points in the company and outside the company where data is generated in the first place and which data is relevant for the company. For example, in connection with customer data, it is possible to consider which contact points exist with customers and what data is generated at the various online and offline contact points. Examples of customer-related data include data on behavior on the company's own website, interaction data in social media, or data generated in the course of product use.

The second step, "Understand," involves an examination and understanding of the data generated. Two aspects are of central importance here:

- To understand how the data came about and what it says in context—and what conclusions cannot be drawn from the data.
- To understand exactly what the analysis results (e.g., a high-density metric) now mean.

In this step, a common understanding about the data used must be created, otherwise the results will be interpreted differently.

Once the data has been collected, evaluated, and understood, the third step is to make informed decisions based on this data. The decisions can be made either by a human or by an algorithm (e.g., an AI). If the decision is made by a human, the data must be prepared in advance (e.g., by means of suitable visualization) in such a way that it can be quickly absorbed and understood by the decision maker.

The fourth process step, "Automate," means firstly that data collection, processing, and visualization are no longer carried out manually, but are supported as far as possible by technology. Secondly, automation also means using technology to produce better results. An example of pure automation of activities is the technology-assisted creation of reports. The use of AI to analyze and evaluate large volumes of data is an example of gaining better insights, since AI can recognize correlations even in large volumes of data that remain hidden to a human.

The "Execute" phase of the process involves the continuous repetition of the four phases mentioned above. In this process, the knowledge gained through the process is continuously communicated to the company. However, this phase is covered in more detail in Chap. 5.

4.2 Collect—Collect Data

To be able to work and decide in a data-driven way, we need a basis. This is formed in the first step of the process by collecting data from various online and offline sources, storing it and preparing it for further processing.

In this chapter, we deal with this,

- Which concrete and comprehensible benefits arise for a company through data-driven marketing.
- What different types and sources of data there are and how we combine the data we obtain.
- What challenges arise from fragmented solutions in working with and using the data, and how we can address these through structural and process measures; and
- What factors we should consider when selecting a technology.

4.2.1 What Is Data?

In order to be able to use data in a goal-oriented manner, it is necessary that we deal with the nature of data. What is data in the first place? What data do we have at our disposal? From which sources does this data originate and in what form is it available? And perhaps most importantly, what data can help us optimize the decisions we need to make every day? These are the questions we will address in the following.

Data are initially nothing more than character strings, i.e., letters, numbers, or symbols. The concrete meaning of a string of characters is not clear at first. Data only becomes meaningful when it is placed in a context. This is how data becomes information. Finally, knowledge results from a linkage of information: The character string 190281 can represent a date. Without context, however, it is not possible to work with this date. But if we now know that this is a birthday, the date becomes information. If we link the date of birth with the knowledge that people are happy to receive gifts on their birthday, and if we as store operators also have a name and an e-mail address for the birthday, then we could send a birthday voucher, for example.

Data are therefore strings of characters that, when placed in a context (see the following example), provide information about an issue. In a marketing-related context, we need data for three reasons:

1. Data is the basis for any marketing measures. Without data, no sales letter can be written and no recommendation system can be designed.
2. Data form the basis for analyses, which can be used to draw conclusions about the effectiveness of the measures taken.
3. Data is the prerequisite for reporting, which informs a user about implemented measures and their effectiveness.

Today, digitization and the resulting low barriers to entry enable almost anyone to access and process data. Digitization makes it easy to collect, store, prepare, process, and transmit data.

> **Example: Relevant Data for a Clothing Store**
> The clothing store distinguishes between offline and online data. Offline data is generated each time a purchase is made in the store: The receipt shows the date, the products purchased, their quantity and price, and the total amount. However, this data cannot easily be assigned to a specific person. Even if this person were to store at the store more often, at the data level they are different people.
> However, if the person uses a customer card, then products and quantities purchased over time can be assigned to that person.
> The situation is similar in the online store: a user can either store as a "guest" or create a user account. Only in the latter case can purchases be assigned to a specific person.
> The example shows very well the importance of context: Viewed on their own, the receipts only say something about sales. Only when they are put into context by assigning them to a person does real added value arise for the company.

One of the key challenges in data-driven marketing is to distinguish data into "relevant" and "not relevant" regardless of its volume, source, or format. This makes the content and message of data the decisive criterion. However, in order to distinguish relevant from irrelevant data, observation, consolidation and interpretation are required. This process is accompanied by continuous learning.

4.2.2 How Can We Differentiate Data?

Data can be differentiated with regard to a number of criteria. A first differentiation distinguishes between quantitative data (e.g., price of a product or size of a garment) and qualitative data (e.g., written product evaluation). Furthermore, a differentiation can be made between real data (e.g., address of a customer) and calculated data (e.g., contribution margin for customer A per year). From another point of view, a differentiation can be made between plaintext data and non-plaintext data.

A fourth approach differentiates between personal, anonymized, and pseudonymized data:

- Personal Data: Personal data refers to an identified or identifiable natural person.[1] This includes both data that can be directly attributed to a person (e.g., a data record with name and address) and data that can be used to draw conclusions about a natural person in a roundabout way (e.g., merging individual data from different files to produce a complete address data record).
- Anonymized data: Anonymized data cannot be assigned to a natural person. This can be data that either had no reference to a natural person at the time of collection or for which the reference was subsequently eliminated.
- Pseudonymized data: This is data that a processor can no longer assign to a natural person. However, assignment rules exist so that re-identification is possible for third parties (e.g., if the names in an address data record are replaced by sequential numbers). The assignment of numbers to names is stored in a separate file. The processor only receives the file with the sequential numbers. Thus, the processor cannot establish a personal reference. However, a third person who has both files and the assignment rule can re-establish the personal reference.

Ultimately, the difference between anonymized and pseudonymized data is that with pseudonymized data, a processor cannot establish a

---

[1] Cf. Art. 4 No. 1 DS-GVO.

personal reference, but third parties can. With anonymized data, it is not possible for anyone to establish an association with a natural person.

Finally, a distinction can also be made between "normal" data and Big Data. As the name implies, we speak of big data when the data volumes are so large that they cannot be analyzed using conventional methods and tools. Other characteristics of Big Data are that it is created at a high speed and that it comes from a variety of different data sources. "Normal" data arises, for example, from the evaluation of web pages, whereas Big Data arises from the merging of data from different sources. For companies, it is now relevant when normal data and when Big Data should or must be used. For this purpose, it is important to clarify which questions should be answered at all. Big data is not always necessary for this, because analyzing it involves more effort.

4.2.3 Which Data from Which Sources Can Be Used?

Different data is relevant for marketing. A company receives the most granular data from the digital area, since the entire customer journey can be measured and thus actively generates data. The website is an important aspect of this: Because not only can data be obtained through it via web analytics tools, but a website also needs data—for example, for personalization. Three types of data are distinguished in the use of data:

1. First-party data: This is data that a company owns itself, e.g., CRM data or transaction data from its own website.
2. Second-party data: Data that a company obtains from third parties is referred to as second-party data. Third parties are, for example, marketing partners such as publishers. The company has access to this data through corresponding agreements.
3. Third-party data: Such data is collected by professional data providers through various methods, usually offered through marketplaces and purchased by advertisers for campaigns.

When collecting data, companies should be guided by the omnichannel concept and collect data across all channels. A distinction can be made between internal and external data. One possible source of internal data is the company's operating processes. During day-to-day business, data can be obtained about the company's customers, but also about other players such as suppliers or partner companies from processes. Specifically, this data arises, for example, from a visit to the website by a customer (e.g., visit durations, pages visited, ads clicked, cookies …) or from transactions (e.g., demographic data, purchasing behavior …). However, this data is not only generated online, but also offline, such as cash register receipts or through the history of a postal communication with the customer.

The Internet and social media represent a data source for external data. The data is not generated by interactions with the company itself. Rather, this data can be viewed publicly by anyone and can be collected, processed and analyzed manually or using suitable tools. This data can be used, for example, to obtain information on the assessment of a company's own products or those of its competitors, as well as on trends and developments. Concrete sources include social media profiles, rating sites, forums, blogs, etc. As already mentioned, data acquired from other companies (second- and third-party data) is another source of data. The scope and content of this data depends on the needs of the purchasing company.

4.2.4 More Data, More Knowledge?

But even though vast amounts of data are generated, more data does not automatically mean more knowledge. Awareness of the added value that comes from making decisions based on sound data has yet to catch on—as does the understanding that it is negligent to make a decision based on hunches or rough guesses.

Making decisions more data-driven can have a number of different causes: For example, that equity investors demand a higher return on investment or that decision makers want greater certainty in their actions. A different performance of the company compared to the rest of the industry can also be a trigger for (stronger) engagement with data-driven marketing.

Admittedly, making data-based decisions is only one component in meeting the three requirements outlined above. But one thing is certain: Those who do not invest in the collection and evaluation of data now will be left behind in the competitive arena later on. A study by the McKinsey Global Institute concluded that companies that consistently rely on data are not only 23 times more likely to acquire new customers and six times more likely to retain them, but are also 19 times more likely to generate profits (see Forbes, 2016).

4.2.5 How Do Data Silos Arise and How Do We Deal with Them?

Silos in the company are a major challenge for data quality. Data silos occur when the individual divisions of a company have their own data collections. Data silos can be formed intentionally[2] or unintentionally. Unintentionally, they are created, among other things, by customers coming into contact with the company at different touchpoints. Different data on the same customer is created at the respective touchpoints. As business units or departments work with different tools, this data is not stored centrally, but in different programs or databases to which only certain user groups have access. Different tools can also be used within an area, for example, if a team uses a special tool only for sending newsletters and stores data in it (e.g., for newsletter unsubscribes). Often, work is done only within this system, i.e., the data is never exported from the tool and merged with the data of other users (intra- or cross-divisional). But this would be important in the newsletter example: because the information that a customer unsubscribes from the newsletter could be an indication of customer churn. And this information should also be available to other teams (e.g., Customer Retention).

In practice, however, the silo problem is evident in both established companies and young companies. In the case of established solutions, silos have often arisen from the fact that the online area was created structurally alongside the offline area in the past and initially used its own solutions

---

[2] Data silos are not bad in themselves. In certain contexts, they are even necessary and are created intentionally, e.g., to meet data protection requirements.

(isolated solutions). In the case of young companies, data silos are often a consequence of rapid growth, i.e., the individual areas quickly sought their own solutions in order to be able to act quickly. However, only the individual area was considered; the overall picture was not taken into account.

In theory, this silo problem—regardless of how it arose—can be easily fixed:

- Organizationally by restructuring the areas. Specifically, this means structurally combining the data owners into one work team.
- Organizationally through the establishment of cross-departmental processes. One example of this would be the holding of regular meetings between the data owners in order to be able to reconcile the data.
- Technically, there are two variants, either implementing only one solution to replace the solutions previously used in the silos, or merging the data stored in the siloed solutions into a third tool.

By breaking down or avoiding silos, data can be merged and the quality of the data ensured. Furthermore, customers can be viewed from different perspectives (e.g., online and offline behavior) and thus better understood. This is also a prerequisite for consistently addressing customers across all touchpoints.

> **Example: Effect and Dissolution of Data Silos**
>
> Our clothing store entered online retail 10 years ago and has set up a separate area for this purpose. This leads to a number of problems, as the following example shows: A customer with a customer account buys exclusively reduced clothing via the website, frequently clicks on "sales" banners and also makes purchases via this. In this respect, the conclusion is obvious that this is a bargain hunter who can be addressed exclusively by low prices and special offers. However, the same customer shows a completely different behavior in stationary retail and also buys high-quality, non-reduced goods. However, this contradictory behavior of the person would not be detected if only the online or only the offline data were evaluated.
>
> However, since the online and offline areas work with different systems that are very specific and whose functions are not covered by a common solution, the decision was made to consolidate the data from the online and offline areas in another solution to which both areas have access—in compliance with data protection-related requirements.

However, in order to build up a holistic picture of the customer, it makes sense to create a use case. Based on this, we can answer the question of what data is required to develop a customer profile. In SteerCo meetings (Steering Committee Meeting), the information requirements can then be compared: What data is required and who has this data? For the example given, the following data can be helpful:

- Surfing behavior of the user on the page
- Purchasing behavior of the user (online and offline)
- Advertisements played (paid advertising, non-paid advertising, newsletters, Facebook ads, Google ads…)

Based on this data, we can perform customer mapping and decide which data will add the most value for us.

4.2.6 What Criteria Are Relevant in the Choice of Technology?

The selection of a technology represents a central decision. Here, we must first decide whether a solution should be developed internally or whether an existing, external solution should be used.

If we decide to use an external provider, on the one hand, an evaluation catalog can be derived from theory (=specifications), against which we can check the tools under consideration. On the other hand, service providers can also support us in our search for the right tool. For example, the consulting firm Gartner regularly conducts analyses of technology providers (=Gartner Magic Quadrant) (see Gartner, 2020a). Furthermore, peer reviews can also help in the selection of software (cf. Gartner, 2020b).

The question of the right technology is easier to answer if our company has clearly defined its objectives, strategy, data protection and data security in advance. Derived from these framework conditions, we can create a catalog of requirements that can be used to determine the optimal technology for our company. In my experience, the most important requirements for a tool are:

- Reputation and location: The reputation of the tool provider and the server location are important decision criteria for projects with sensitive data, especially with regard to data protection regulations within the EU. In addition, the possibility of dedicated support during the project is an important factor.
- Data collection and storage: When collecting and storing data, both the format in which the data is collected and stored and the location where it is stored are important factors. It must be ensured that the data can subsequently be accessed and exported in the required formats.
- Third-party data integration: It is important that vendors integrate various third-party data providers on the platform and that their tools can be purchased through a marketplace.
- Customer segmentation: In customer segmentation, anonymous customers are identified and assigned to a defined segment. Segment assignment can be rule-based or based on complex statistical methods.
- Distribution of segments: This requirement describes the ability of the DMP to export segments to other, external systems. The interfaces and real-time capability of the DMP are crucial here.
- Reporting and Analysis: This requirement identifies the existing reporting and analysis functionality of the DMP.

Based on the prepared requirements catalog, the shortlisted service providers can be evaluated so that a rating system can be created for each technology solution. This allows companies to ensure that an objective evaluation of the possible scenarios takes place. However, it should be noted that it is very difficult to obtain a comprehensive overview of the market. For this reason, experts should be consulted who can narrow down the selection of tools in advance based on the requirements that have been developed.

4.2.7 What General Conditions Do We Have to Consider?

When collecting and working with data, it is imperative that legal requirements are taken into account. This means that companies must find processes and solutions that demonstrate compliance with legal requirements. The following questions are often at the center of this, and they make the company's responsibility in dealing with data clearly visible:

- Why do we process data?
- How do we process data?
- With what legal basis do we process the data?
- How safe are our processing routes?
- Are we allowed to process the data at all?
- In which way does the user give us his consent, if necessary and required?

Current topics of particular relevance for companies in online marketing include the General Data Protection Regulation (GDPR) and related topics such as eprivacy and consent management. For example, the DSGVO forces to avoid third-party cookies, i.e., cookies that are not set by the website itself but by a third party, as much as possible and to solve as much as possible via first-party cookies or technologies. This means that a website operator may collect and process data for itself, but may not pass it on to third parties. In this context, it is the task of the company to react to such developments and, for example, to align its own technologies accordingly. However, the marketer should always keep in mind the effort required for collection and, if necessary, refrain from collecting data.

Finally, it should be noted that valid data cannot be collected for every situation. Often the data is missing completely, in other cases, it is only available in fragments. There is therefore a certain degree of uncertainty when a decision has to be made in this situation. However, it is also not always necessary to have complete information. But before we ignore this (incomplete) data and know nothing at all, it is better to use it and at least get an overview and a rough guide.

4.2.8 Guiding Questions for Collect

- What questions should be answered by data?
- How quickly should the questions be answered?
- What data do we need to answer these questions in a timely manner?
- What sources do we have available to answer the questions?

- Which of the relevant data are already available to us? How is the data available (format, real-time or near-time, API)?
- Do we already have access to the data (e.g., access from a legal perspective)?
- Do those people/areas that should have access actually have the ability to retrieve data?
- Have contractual issues been clarified (e.g., legal consent to collect the data)?
- Have the data already been checked for completeness and correctness?
- If data is not currently available: How quickly can we get the data?

4.3 Understand—Understanding the Collected Data

> It's better to be approximately right than precisely wrong.
> (John Maynard Keynes, British economist, 1883–1946)

After we have collected data from different sources in the first step, the second step is to understand this data. Understanding means, on the one hand, being able to comprehend how the data came about, how the individual data influence each other, what the most important drivers are and, above all, being able to comprehend what the data say in their respective context. On the other hand, understanding also means being able to grasp the results of the analysis and to know what, for example, a specific, highly condensed key figure actually says.

In this chapter, we deal with this,

- Reasons why understanding is important and what cognitive skills an analyst needs to understand data.
- Which technical measures are the prerequisite for being able to understand data.
- What simple measures there are to make data accessible to oneself.
- How we can continue to work with the data we collect to produce meaningful analysis.

4.3.1 Why Is Understanding Central?

One of the most important aspects for decision makers in companies is to understand the context as well as the consequences of decisions. In essence, this is about the correct interpretation of data in order to create added value for the company, for example, by realizing previously unused potential or reducing costs.

It would be grossly negligent to make decisions solely on the basis of a hunch or a rough guess. This is especially true in the online space, where the granularity of data has increased dramatically: Whereas a few years ago it was perfectly sufficient to get an overview of the status quo using dashboards, today analyzing individual drivers leads to more targeted and effective results. One of the reasons for this is that there is much more data available today. In concrete terms, this means that we can analyze much better how a result value came about and which causes are responsible for this. For marketing, this in turn means that more data can be generated and used along the entire customer journey. In particular, it is important to be able to activate the collected data later, i.e., to extract added value from it.

> **Example: Why an Understanding Is Important**
>
> The head of marketing at the clothing company would like to decide at the beginning of the year how to allocate his budget to the individual channels. He uses the sales per channel as a decision criterion, i.e., channels with a higher share of total sales also receive more budget. To do this, he looks at the channel attribution, i.e., the distribution of sales and revenue per channel. By default, this is reported on "Last click." This means that the channel on which the customer had his last contact is attributed to the sale. On this basis, and taking into account the reports he always receives from his staff, the head of marketing concludes to invest the entire budget in Google Ads. However, an employee points out to him that this is not necessarily the right decision: if you look at the allocation differently, for example, and divide the sale evenly between the channels that led to the purchase, a completely different picture emerges. It then becomes apparent that the number of clicks on the Google Ads ads and sales was always high when the company sent customer mailings by post. This creates a completely different picture, namely that the mailing is also very important. If the mailing was deprioritized and the entire in budget was invested in Google Ads, a significant drop in sales would probably be seen after a few weeks or months.

And in my opinion, this is also the big difference between data-driven marketing and market research: In market research, we work with models that have been generated on the basis of empirical studies or theoretical considerations and claim to have general validity. However, these models have to be interpreted for application in a specific situation. Data-driven marketing is different: Here, precise data is generated, on the basis of which decisions can be made or these can be returned to their intended use (data activation).

4.3.2 What Conditions Do We Need to Be Able to Understand?

In order to be able to make data-driven decisions, both human and technical prerequisites are necessary. The technical prerequisites are the same as the maturity level of the company and relate, for example, to the tools available. In combination with the analytical skills of a human being, data-driven decisions can then be made.

4.3.2.1 Technical Requirements

Many projects have already shown me that traditional companies tend to invest less in technical requirements. The reason for this is that, on the one hand, the benefits of investments are often difficult to measure and, on the other, there is a fear of being replaced by technology.

A good example to illustrate the technical prerequisite and its effect is a simple sheet of paper: Let us try to remember five different numbers and calculate the average value from them. This calculation is very difficult to do in the head. With a simple sheet of paper and a pencil, you create completely different conditions.

Pen and paper, figuratively speaking, must also be procured in the company. This creates the prerequisite for processing the volumes of data available. The issue of isolated/silo solutions and joint, cross-divisional solutions also plays a role at this point. If all relevant business questions can be answered with isolated solutions and a holistic view of the customer is not relevant, the typically existing technical requirements may be sufficient. However, if a holistic view of the customer is required or necessary, the structures and processes already described in Sect. 2.5 (breaking down silos) must be created.

4.3.2.2 Analytical Requirements

The analytical requirements relate to the person himself. As a manager, I like to ask people to complete so-called brainteasers during the application process. In addition to the human values queried in other ways, these show me whether the person fits into my team. The brainteasers enable me to determine, for example, whether the person has an idea of how data is created.

Analytical prerequisites should also be present in the people who prepare decisions or make the decisions themselves. It is also necessary for a decision maker to know how data is created and processed in order to be able to "read" and interpret data correctly.

To be able to implement an analysis, the following additional skills are needed in my view:

- Problem understanding, i.e., being able to cognitively grasp what challenge or task needs to be solved
- Decomposing the problem into components, since problems are often very complicated or even complex
- Structuring the sub-aspects and finding solutions for these sub-aspects
- Consider the interactions between the partial solutions
- Derive a strategy and solution to the overall problem

4.3.3 What Must a Technical Preparation Look Like?

The technical preparation of data is often referred to as ETL processes (Extract, Transform, and Load processes, see also the following explanation). ETL therefore means extracting the required data from the respective data sources and preparing it. ETL processes are necessary because data is usually redundant or has completely different structures. The technical preparation ultimately ensures the quality and consistency of the data.

This process is mostly used for larger amounts of data (Big Data), as the preparation for small amounts of data/sources is usually not in relation to the achievable output.

> **Explanation: What Does ETL Mean?**
> - Extract: Export or extract the data sources identified as relevant in Chap. 2, including data subsets.
> - Transform: Unification of data structures, formats, and schemas. This is often the case because data from different systems cannot be superimposed due to different structures. Thus, in this phase, the formatting is standardized and possible errors such as identical information (duplicates) are avoided. Subsequently, the data is aggregated again, if necessary, and transferred to a target format.
> - Load: The final step is loading the prepared data into the target system. The data is also physically transferred to the target system.

Depending on whether online or offline data is to be prepared, different requirements arise in terms of speed and latency, since the data may have to be activated near-time.

Possible application areas for ETL processes are

- Merging data from diverse sources
- Storage in a data warehouse (DWH)
- Data preparation for business intelligence (BI) applications such as visualization

4.3.4 How Can We Tap into Data?

Perhaps the most important aspect related to data is understanding it. In most cases, the challenge is not that the analyst himself understands his data. Rather, it is the analyst's task to make the data understandable to the target group (i.e., decision makers such as department heads, management ...). This makes the preparation and presentation of data and insights an important task for the analyst. Because only when the decision makers have understood the core statements can they also take them into account in their decision-making.

However, bringing out the insights to decision makers is a challenge for many analysts. It is very difficult to create presentations that are not a mere stringing together of facts and figures and confuse the listener by their scale. And as noted in Chap. 2, the amount of data available increases—and the more data the analyst has available, the higher the risk that a presentation based on it will get lost in the numbers. Even if a target group can intellectually grasp and understand the data presented, this does not mean that the numbers will also inspire them to act. Understanding data is very important, especially for decision makers, because questioning tends to decrease with the hierarchical level of the decision maker: If the decision maker does not understand something, he does not give himself the nerve to question. In case of doubt, he simply decides against a sensible solution because he cannot understand the underlying reasoning and therefore rejects it.

4.3.5 What Does Emotionalizing Data Mean?

If, as an analyst, you come across data that requires immediate action, then this urgency must also reach the decision makers. You must achieve that the decision makers not only understand the numbers, but also recognize the seriousness of the situation and feel compelled to take immediate action. That is why a presentation must reach decision makers on an emotional level.

4.3.6 How Can We Facilitate an Understanding?

People can grasp facts more easily if they refer to objects or facts that are familiar to them. This is especially true for grasping and understanding numbers. Numbers should always be placed in relation to something familiar. In this way, the analyst, or the target group of the analysis (e.g., division management, executive board, etc.), does not have to rack their brains over how large or small a number is. Especially complex numbers can be captured much easier this way. Ultimately, the analyst has to place the figures to be presented on a scale. This can involve a reference to a known size or distance, a known time segment, or a speed. In the following, I will present some examples.

4.3.6.1 Reference to a Comparable Size

Data can be linked to a comparable quantity, e.g., via a length, height, thickness, or distance. However, this comparison is only useful if the target audience is aware of the comparable quantity. For example, "If all XY were placed end to end, they would reach 1.5 times around the earth."

4.3.6.2 Establishing a Time Reference

Another possibility is to establish a reference via time. It is measured in seconds, minutes, hours, days, etc. However, this can sometimes be difficult to imagine for a target group. A time reference can be compared, for example, via flight times between well-known cities, the duration of an episode of a well-known sitcom, or the time it takes to cook pasta.

4.3.6.3 Reference to Known Objects

However, a reference can be established not only via quantities or a time reference, but also via known objects. Objects can be, for example, people or places. Let us assume one million users who have visited a website in a period of 1 month. To make the number of one million people more tangible, a comparison can be made with the number of visitors to a stadium. The FC Bayern Munich stadium can accommodate around 75,000 people (=sum of seats, standing room, business, and box seats). A user number of one million people per month therefore means that the Allianz Arena could be filled almost 13.5 times.

A comparison also works to illustrate sums of money: Jeff Bezos is currently the richest man on earth with an estimated fortune of around $130 billion.[3] There are currently around 570,000 homeless people living in the USA. The purchase of a house in the USA costs around 200,000 US dollars. One can now make the following calculation: If Jeff Bezos were to buy a house for every homeless American, he would still have 16 billion US dollars left, or around 12% of his current fortune.

The comparisons and illustrations provided can help an analyst make clear to his or her audience the magnitude of an opportunity or even a risk and persuade decision makers to make a decision.

4.3.7 Guiding Questions for Understand

- Are there small or large amounts of data?
- For small amounts of data: How must the data be prepared (e.g., Excel?)? Are the necessary requirements met by the processor?
- For large amounts of data: What are the technical requirements? Which specific software solutions or combinations of software solutions do we need? Is the data technically prepared in such a way that it can be further processed automatically?
- Regardless of the amount of data: Is the data prepared in such a way that a human being can understand it? Is the data prepared in such a way that a decision can be made on the basis of this data?

4.4 Decide—Decide on the Basis of the Collected Data

> Whoever says A does not have to say B, he can also realize that A was wrong. (Bertolt Brecht)

The third phase, "Decide," refers to the making of decisions. However, in this phase we do not only want to look at the statement of intent of a

---

[3] Cf. finanzen.net (2020), n.d.

decision maker, but also at what contribution an analyst can make to making a good decision.

In this chapter, we deal with this,

- What differentiates data-driven decisions from gut decisions.
- What different types of decisions have to be made in a company.
- What data a decision maker must have in order to make a good decision.
- Which possibilities exist to prepare and visualize data for a decision maker.

4.4.1 What Distinguishes a Data-Driven Decision from a Gut Decision?

Every manager, every person in leadership responsibility, makes a multitude of decisions every day. Many of these decisions are intuitive or influenced by emotions or a gut feeling. Other decisions are made automatically by people, i.e., they do not think about the decision but make it out of habit. In marketing, this is also referred to as habitualized decisions. The decision in favor of a brand when buying toilet paper is usually such a habitualized decision. Very few decisions in everyday professional life are made on the basis of data and thus on a well-founded basis.

But let us be honest: On what basis a decision was made that in retrospect turns out to be "right," "good," or "beneficial" is ultimately completely irrelevant. But: Making good decisions is one of the central characteristics of a manager. So if a manager makes one or more bad decisions, it is better to be able to refer to data and not have to admit to having made an (unfortunately wrong) gut decision.

Gut decision vs. data-driven decision, that sounds like the blatant difference between Mr. Spock and Captain Kirk: Mr. Spock, the passionless and emotionless Vulcan, who can justify every decision he makes with his logic. Captain Kirk behaves quite differently, listening to his first officer's opinion, but often not following it. He regularly makes his decisions "on instinct."

Gut decisions are made intuitively, without much thought, and therefore very quickly. Human intuition is based on past experience and emotions. The decision situation is simplified considerably by not taking into account many factors influencing the situation. If a person wanted to take all influencing factors into account for every decision, he would hardly be able to make a decision at all.

In contrast, data-driven decisions require a comprehensive examination of the current situation. The decision maker tries to understand the situation as best as possible using the available data. The decision maker sifts through the facts and weighs them up. If the decision situation is well structured, then influencing variables and their interdependencies as well as the main drivers and their influence on the outcome variable or variables are known. The decision maker can understand these hypotheses about reality, weigh them, and then make a decision. Key performance indicators (KPIs) can also be used to make data-based decisions. KPIs represent an aggregation of various drivers which, as key performance indicators, provide information about specific business facts. KPIs therefore translate more complex issues into a very simple indicator. KPIs must always be considered on a company-specific basis and, above all, must remain comparable over time.

However, not only the decision-making situation itself and the decision-making process are different for the two variants, but also the monitoring and controlling of the decision: In the case of a gut decision, there are few opportunities to follow up, because either the situation develops as predicted or it develops differently. With data-driven decisions, on the other hand, there is the possibility of tracking and learning: For example, if a driver tree is set up with a top metric, the decision maker can immediately see a trend and determine whether the assumed relationships are correct. If this is not the case, the hypothesis framework must be modified and optimized. Ultimately, data proves the path taken. This does not work with gut decisions; here, the decision maker can only recognize much later whether he was right.

4.4.2 What Types of Decisions Are Made in Companies?

The decisions to be made in a company can be differentiated in terms of a number of dimensions. One of these dimensions is the question of who makes the decision. At this point, we want to distinguish between three levels: the employee level, middle management, and the executive or board level. The decisions to be made by the employee have an exclusively operational character. He often only has to choose between different alternatives. The greatest challenge at this level can be seen in the frequency with which the person in charge has to make decisions. Middle management cannot start from predefined alternatives, but the solutions are within a limited radius. What is particularly difficult for this level is the fact that the individual decisions often cannot be viewed in isolation from one another and exhibit interdependencies. And what is the situation at the management or board level? The decisions to be made are of a strategic nature, i.e., they have a long-term effect and are also difficult to revise in many cases. It is therefore all the more important to make "good" decisions. Decision makers at this level are confronted with very unstructured decision-making situations and a high degree of uncertainty.

The type and amount of information available to a decision maker also differs. The decisions that a department head has to make are usually much more granular than at the board level. This means that a department head has much more information at his disposal; he has a completely different depth of information. The task of the department head is now to prepare the available data in such a way that the board level can make a decision. To do this, he or she condenses the information (e.g., in the form of KPIs), which is then used by the board level to make a decision. The employee must also ensure that the board level trusts the data on which the decision will then be based.

4.4.3 What Are the Requirements for Making a Good Decision?

I have learned in my career that making sound decisions is based on three assumptions. But beware:

> An informed decision is not necessarily the right decision.

The three prerequisites for making an informed decision are

- The data or information basis on which the decision is based.
- The analytical skills available to assess the data.
- The time available for making the decision.

But why do managers sometimes find it difficult to make "good" decisions if all they had to do in their decision-making was consider the three points listed?

Management is often "under power" and managers therefore do not have the opportunity to consider all the data necessary for making a good decision. The context in which managers make their decisions therefore often makes it impossible to take the three aspects listed fully or even only partially into account—even if this would be desirable. This is because digitization in particular is presenting managers with ever greater challenges, tasks, and decisions (see the following explanation).

> **Explanation: Old vs. New Economy**
>
> In an industrial company in the 1980s, the responsibilities of employees and managers were clearly distributed and delineated: There was a shift supervisor in production who controlled his area and who was responsible for all measures and incidents in his area. If a loss occurred, it only affected his own area. And the shift supervisor also had an appropriate time frame available to rectify the problem.
>
> In a modern company, the situation is different because the issue of scaling is playing an increasingly important role. As a result, managers are taking on responsibility for more areas in which many processes or sub-processes are automated and there is no manual control. As a result, it can quickly happen that managers have to bear much greater responsibility. In addition, there is less time available for the complex decisions they have to make. In plain language, this means that there is less time available for more difficult decision-making situations and, in addition, the decisions have a greater impact.

4.4.4 What Role Does the Time Factor Play in Decisions?

Time plays a role in decision-making in that the decision maker can consider more data or information for decision-making if he has more time.

The less time he has, the more reactive the person is and the higher the maturity level of the technology must be in order to be able to operate successfully on the market. However, this also means that the respective decision-making bases must be available to the decision maker in aggregated form. If the necessary time is not available, the decision maker will probably rely exclusively on his intuition, even though this may be wrong. If it is therefore possible to run through scenarios of the decisions in terms of time, it is easier to make a decision. In addition, it is possible to assess which of the available alternatives is the most advantageous for the company.

4.4.5 How Can We Visualize Data?

An important aspect of generating understanding for data is its visualization, i.e., the graphical representation of data. After all, mountains of data that lie hidden on hard drives do not create any added value. Only visualizations help to derive recommendations for action and to uncover optimization potential. Because visualizations can be used to

- Support the human ability to absorb by making targeted use of shapes, colors, and patterns.
- Data presented in an easily understandable way.
- Trends, developments, outliers, correlations between developments, etc., can be identified more quickly.
- Even non-specialists can recognize the key messages of data more easily.

A visualization can be done, e.g., by diagrams, maps, or graphics. With the help of these visualization forms, it is very easy to recognize trends as well as outliers in large amounts of data. If colors are also used, core statements can be made even clearer. After all, people are used to absorbing data visually.

However, good data visualization is a balancing act for the analyst: He must always balance between the two dimensions of "form" and "function." For example, a simple graphic may not attract enough attention to convey the information that is important to the audience. On the other hand, a laboriously prepared and extensive graphic may be completely unsuitable for conveying the right message to the man or woman. Ultimately, the analyst must succeed in achieving a symbiosis between numerical material and pictorial presentation.

In general, when visualizing data, both the prior knowledge and background of the addressees should be taken into account (to whom am I presenting the data?) and the data itself should be put into context (e.g., use titles for charts, indicate units on axes, add background information if available).

In the best case, it is possible to tell stories with data. To do this, complex facts must be represented in a simple way. Rudyard Kipling, a British writer and poet who died almost 100 years ago, once formulated:

> If history were taught in the form of stories, it would never be forgotten.

The same is essentially true for data: The addressees of presentations remember data more easily if it is packaged in an (exciting) story. But how do I tell an exciting story?

Regardless of the data itself, I think a story should have the following plot:

- Main characters,
- Problem, and
- Happy ending

In our story, the KPI takes on the role of the main character. The problem is a business challenge and a solution based on this data is the happy ending. Depending on the relationship between the storyteller (analyst) and the listener (decision maker), the introduction to the story should describe the initial situation. This can include, for example, what methods were used to clean up the situation. Next, possible solutions should be pointed out and supported by associated advantages and disadvantages. This usually creates the basis for decision-making, on the basis of which the listener can determine the end of the story.

4.4.6 Data Versus Gut—Or Better in Combination?

However, what has been said so far is not meant to sing the praises of data-informed decisions. Challenges can arise even with data-based decisions. For example, in my career, I have often found that the term KPI is greatly misinterpreted: For example, as external consultants on projects, we have more than once been given lists of around 100 KPIs, all of which were of course important to the company. However, this high number of indicators contradicts the basic idea of KPIs—especially since the word itself, through the component "key" or "key," gives an indication that only the very most important indicators can be KPIs.

In some projects, we set the reports to "zero," i.e., we replaced all the key figures and KPIs they usually contained with zeros. In many cases, however, the recipients of the reports did not even notice this because they were so overwhelmed with reports that they could only read a small part of them, let alone understand them. Thus, decisions were made, but not on the basis of reports or data. In other words, too much data can have the opposite effect of what was originally intended. Or to put it a bit more bluntly: Instead of using the data to make decisions, the over-supply of data ensured that fewer decisions were made on the basis of data.

In this respect, a combination of gut decision-making and data foundation could be a suitable way forward. We cannot therefore assume that decisions will only be made on the basis of facts, as this could also end in disaster. In the end, at least for the time being, people will always make decisions, so emotions or feelings cannot be ruled out. But: The basis on which decisions are made can be significantly improved. In concrete terms, this means that data is necessary for a decision, since decisions cannot be made in a vacuum. The decision maker can use data to get a picture of the situation, to inform himself. At the same time, however, they should not disregard their experience, intuition, and gut feeling when making a decision. This is all the more true if the available data is contradictory or if there is too much or too little data. Or, to put it another way, we need to get Mr. Spock and Captain Kirk to work together in a better, more goal-oriented way.

4.4.7 Guiding Questions for Decide

- Where do we currently stand with our business? Are we already making decisions based on data or are we still relying on gut decisions? Can we back up decisions with numbers?
- If a result was good or bad, can we prove why?
- Can we already make predictions?
- How quickly do we need to make decisions? Is the data available in such a way that we can make a decision sufficiently quickly?
- What do we need to do to make (more) data-informed decisions in the future?

4.5 Automate—Automation

The fundamental question of this chapter is "What can we automate and in what way?" Automation should be considered on two levels: First, automation that produces better results than before, e.g., through the use of artificial intelligence (AI). And secondly, an automation of processes that were previously carried out manually, thereby both increasing speed and reducing the error rate. In this chapter, we take a closer look at this,

- Why automation is inevitable.
- Which prerequisites must be met for automation.
- What forms of automation there are.
- What benefits result from automation.
- What counterarguments there are to automation.
- Which dangers can result from digitization.

4.5.1 Why Can't We Get Around Automation?

In my opinion, automation in marketing (=marketing automation) is inevitable. Data is the basis for making good marketing decisions. In this context, "good" means, for example, being able to address the customer at the right time with the right content on the right channel. However, in order to make good decisions, a large amount of data from different sources must be evaluated and analyzed. Put simply, the more data we have, the better decisions we can make. However, data analysts can no longer handle the daily volume of data and information with pen and paper or Excel spreadsheets alone. In concrete terms, this means that even the best marketing manager will not be able to avoid automating processes in the future. It is imperative to automate both decision-making processes and performance reviews. To achieve this, the best company-specific selection of tools must be put together in marketing, and people must be trained to work with these tools. If necessary, standardized tools must also be adapted to a company's specific needs. And automation provides another advantage: machines are more scalable, unlike human labor. So if the amount of data to be processed doubles, this can be absorbed more easily via tools than by hiring new employees.

4.5.2 What Are the Technical Requirements for Automation?

The parameters specified in Chaps. 2 and 3 are the prerequisite for the automation of Big Data solutions, because such a large amount of data can only be processed with a given structure. Without such a structure, no solutions, such as R or Python, can be used. These solutions are used to process large amounts of data. In addition to company-specific solutions, different manufacturers also offer solutions in the automation sector.

A second question in connection with automation is that of the tools to be used. This is because they are at different stages of development and progress. A rough distinction can be made between two approaches. The first variant involves the use of special solutions, i.e., tools that have been developed specifically for a company or are at least adaptable. For example, there are solutions for specific industries that can be adjusted to the specific needs of an individual company. The second variant comprises automation tools that provide standard solutions for a wide range of tasks in marketing and sales. State-of-the-art solutions use a cloud to provide standardized automation solutions for companies of any size.

4.5.3 What Added Value Does AI Create in the Context of Automation?

The first form of automation is aimed at activating previously unused data in order to gain better insights based on this data. AI is used in this form of automation. AI is understood as the ability of a system to imitate human behavior. Consequently, AI-based systems are able to perform tasks that were previously performed by a human. A typical task for an AI is to recognize environmental factors and framework conditions and, based on this, to make decisions aimed at achieving a specific goal. AI is also the generic term for a number of different approaches to imitating human behavior. This includes machine learning, among other things. Machine learning (ML) involves algorithms that can learn from data provided and derive generally applicable rules. The rules later enable the algorithm to interpret new data and make decisions. The more data available to the system for learning, the better the decisions.

When using AI in marketing, a distinction is made between two variants. In the first variant, AI is a fixed component of a solution that a company uses, such as in Google Analytics or Google Ads. Both tools offer intelligent evaluations that support the analyst. This support is helpful even for small websites, because even there, significant amounts of data are already generated that a single analyst could never evaluate on his or her own. In my opinion, Google Analytics, for example, has found a good way. Because with Analytics Intelligence, Alphabet has developed its own AI, which is able to automate the analysis skills of analysts across all data. In my experience, an analyst always looks at the driver trees that are relevant to him. But an AI can now detect changes in the driver trees much more quickly than an analyst, for example.

> **Example: Use of AI in Google Analytics**
>
> Our fashion store uses the Google Analytics tool for evaluation and has set an intelligent alert there. Through this alert, the marketing manager recognizes that the number of search queries for socks has increased. Based on this information, he would come to the conclusion to purchase and stock more socks (more variants, more socks per variant) and also to put more content about socks on the site.
>
> In addition, however, the marketing manager uses an AI that runs through the online store. However, this AI uses not only the search queries, but also the actual sales. This AI also finds that demand has increased in the socks product category. However, the increase in sales only concerns a very specific category of socks (sports socks with compression content). So all measures should focus on this category of socks.
>
> Ultimately, the marketing manager only arrives at the right conclusion by merging two alerts triggered by AIs. He probably would not have reached this conclusion himself due to the large amount of data.

AI will democratize analytics, in my opinion: Because complex prediction algorithms and recommendation engines, which were previously only available to the big players like Google or Amazon, can now be rented.

In the second variant of AI use in data-driven marketing, the available data from different sources is first brought together, e.g., online and offline sources. Then, an AI is used to identify correlations in data and gain insights. The distinctive feature of this approach is that questions are not defined in advance for which the AI should search for an answer in the data. Rather, the AI searches independently for anomalies, e.g., in the form of patterns.

4.5.4 Is Automation Even More Than AI?

As the previous explanations have shown, automation can improve the quality of results because, for example, more data can be taken into account or an AI recognizes patterns that a human would never have identified. A second form of automation is not aimed at obtaining better results, but at having frequently recurring processes performed by a machine rather than a human in the future. Not only can this reduce the time required for execution, but at the same time, human errors can be avoided. After all, it is in the nature of humans to make mistakes. No matter how intelligent a person is, and no matter how many people put their heads together. Errors arise from manual processes such as working with Excel files, other documents or reporting. However, individual, small errors can turn into processes with large errors that are no longer traceable due to reporting and decisions made on the basis of these errors. These sources of error can be avoided through automation.

In data-driven marketing, process automation is also referred to as marketing automation. Marketing automation means that actions are executed automatically based on the data of a CRM system, for example. For example, a customer receives a birthday mailing or personalized offers based on his purchase history.

4.5.5 How Do We Manage to Transfer Our Findings into Processes in an Automated Way?

A third way to automate in data-driven marketing is to automate insights gained through automation (e.g., using AI on merged data) into processes.

Let us take the example of a recommendation engine, i.e., an approach in which individual recommendations are made to a person. In the conventional approach, we would provide different variants of, for example, an ad for socks (variant A, variant B and variant C) to subsets of users and use significance tests to find out which of variants A, B, or C performs best. This can be measured by the conversion rate, i.e., those people who were shown ad A have a significantly larger shopping cart than those users who were shown ads B or C. With the help of an AI, however, it would now be possible to find out whether there is a specific customer segment for which ad B or ad C works better than ad A. For example, ad A may have the best performance (size of shopping cart) across all customers. However, if we only look at customers under 30 years of age, ad B shows the highest performance.

The knowledge gained by an AI that ad B performs best with customers under 30, but ad A performs best with all other customers, can be automated in a further step, i.e., the ads for socks will be alluded to according to age in the future. The objective is therefore to automatically convert the website for 1:1 or 1:n communication.

4.5.6 What Can Be the Causes of Resistance to the Data-Driven Organization?

But there is no light without shadow: Even though the previous explanations have made it clear that automation has a lot of potential for data-driven marketing, there are still reservations about the increasing use of machines. This is not only due to the fact that AI can also be used for illegitimate purposes (see the following explanation). Rather, some people feel uncomfortable when decisions are made by a machine. Especially since the decision cannot always be verified by the human. This can happen especially when using Big Data applications, since decisions are not based on simple if-then logics.

> **Explanation: Cambridge Analytica**
>
> A good example of how data should not be handled is the US election campaign of 2016. It is assumed that the Republican campaign team was able to use the data of around 87 million Facebook users for advertising purposes. The starting point was a third-party mobile application (this is your digital life) that allowed Facebook users to take a personality test. The developer thus possessed detailed information from 270,000 people—both from the questions answered in the application and from the data stored in the people's Facebook profile. Crucially, the company Cambridge Analytica subsequently carried out a comparison between the data from the personality test and the data from profile data. This made it possible to infer the personalities of other people from their Facebook data. Donald Trump's campaign team used this information to target those people whose profile suggested they were undecided about the upcoming election (see Revell, 2018).

In summary, we can state that the goal must be to get a handle on the data. The analysis capabilities of a human being are limited, as only a certain amount of data can be taken into account and overviewed. Due to the amount of data generated today, it is imperative that the analyst be supported by a machine. In the future, there will be no way around this, also due to the constantly increasing data volumes. In the following chapter, we will also discuss how automation can be implemented in the company and how resistance can be overcome.

4.5.7 Guiding Questions for Automate

- Do we need automation? Which process steps are to be automated and to what extent?
- What can we automate to get better or faster at the Collect, Understand, or Decide phases?
- Are the technical prerequisites for automation in place?
- If the requirements are not met: What solutions/software solutions do we need?
- What does automation cost? Does this investment make economic sense in the long term?

4.6 Summary

The implementation of data-driven marketing must always be a company-specific decision. In the case of large companies, it may also make sense to find an individual solution for each business unit. After all, not every company or every business unit necessarily needs AI or extensive software solutions to make decisions.

Company- or business unit-specific also means that copying solutions from other companies does not make sense: It is possible to be inspired and adopt individual ideas. But firstly, the topic is still very new for most companies, and the "holy grail" in the sense of an absolute best solution does not exist, and secondly, a solution that delivers good results for company A does not necessarily have to work for company B as well.

References

Finanzen.net. (2020). *The richest man in the world: This is what Jeff Bezos could buy with his billion-dollar fortune.* finanzen.net. Retrieved May 13, 2020, from https://www.finanzen.net/nachricht/geld-karriere-lifestyle/luxus-einkaeufe-der-reichste-mann-der-welt-das-koennte-jeff-bezos-mit-seinem-milliarden-vermoegen-kaufen-8819897

Forbes. (2016). *Becoming a data driven organization*. Forbes. Retrieved May 13, 2020, from https://www.forbes.com/sites/adigaskell/2016/10/28/becoming-a-data-driven-organization/

Gartner. (2020a). *Magic quadrant research methodology*. Gartner. Retrieved May 13, 2020, from https://www.gartner.com/en/research/methodologies/magic-quadrants-research

Gartner. (2020b). *Enterprise IT software reviews | Gartner peer insights*. Gartner. Retrieved May 13, 2020, from https://www.gartner.com/reviews/home

Revell, T. (2018). How Facebook let a friend pass my data to Cambridge Analytica. *New Scientist*. Retrieved May 13, 2020, from https://www.newscientist.com/article/2166435-how-facebook-let-a-friend-pass-my-data-to-cambridge-analytica/

5

Process Model for Implementing the Data-Driven Organization

Abstract This chapter represents the core of the book by describing the process model step by step.

5.1 Overview

In my opinion, the disadvantage of strategy development procedures that are described down to the smallest detail is that they become outdated very quickly: After all, nothing is more ephemeral than a technology that is hyped today. Technical knowledge also becomes obsolete very quickly. That's why in this chapter, I will show a very generically formulated process. This process will still be valid in the future, regardless of technological developments and other influencing factors. When implementing it, you should keep the Pareto principle in mind: The important thing is to get into practice and into action and not try to come up with a theoretical strategy that is 100% perfect on paper. Ultimately, we work here with a lean approach: think briefly, plan and then do. Then subsequently review, analyze, get feedback, and go into the next iteration loop. We can use various instruments from the lean toolbox: For example, the minimum viable product (MVP) approach when developing internal services for the specialist departments or the fast fail approach when testing new processes.

This approach saves us months of planning processes that result in a pile of paper—which then remains in the drawer but is never or only rudimentarily implemented.

I will explain the seven steps mentioned in Fig. 5.1 in more detail in the course of this chapter. At this point, however, I would like to give you the overall context.

The process begins with the development of the data strategy. This includes conducting an internal and external analysis to determine both the status quo of one's own company and that of one's own industry and the competition. This is followed by the definition of data goals, which are derived from the company's objectives. Data goals answer the

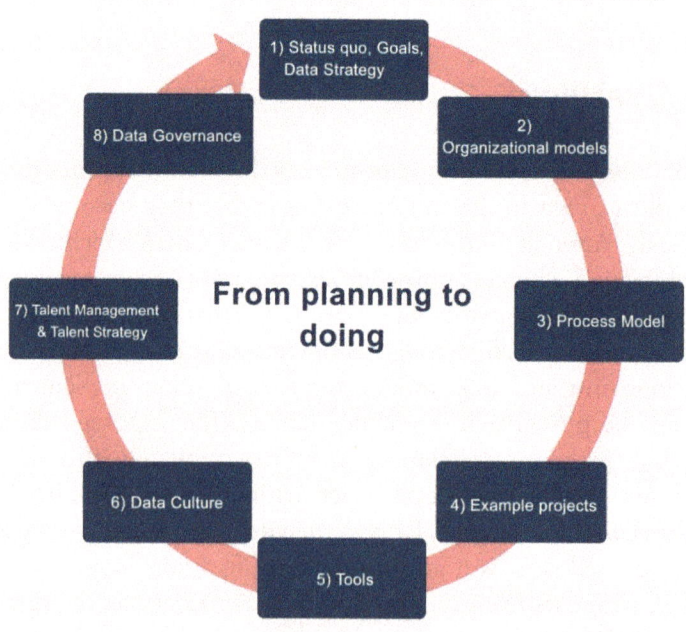

Fig. 5.1 Process for implementing the data-driven organization (Source: Own representation)

question of how data can support the achievement of corporate goals. Subsequently, the concrete data strategy is developed. Here, I address explicit content such as make-or-buy decisions or the question of customer access. The first step concludes with the question of integrating and interacting with stakeholders and ensuring continuous innovation in the area of data & analytics.

The second step in the process involves defining a structure in accordance with the tried-and-tested principle of "structure follows strategy." The task in this step is to find an organizational structure that allows the best possible implementation of the data strategy. For this purpose, I will present three possible variants. After the structure, in the third step, we deal with the process for data and analytics. I show how, in a structured approach, data can be collected, prepared for further processing, a uniform understanding of the data can be built up, decisions can be made based on this data, and the process can be automated as well as anchored in the organization. The data and analytics process is implemented through sample projects. After the strategy, the structures and the processes have been illuminated, the fifth step is to deal with the systems in the form of tools. The last two process steps deal with the question of designing a data-friendly culture and data governance.

5.2 Status Quo, Goals, and Data Strategy

5.2.1 Internal and External Analysis

Before we can start creating a data strategy, we need to determine our status quo and goals, as well as identify the need for action.

Let us start with the analysis to determine the maturity level of the company. The maturity level is important in that many companies already have a data strategy in some form or at least have parts of a strategy in place. In these cases, it would not make sense to start planning from scratch. Rather, we want to build and build on what we already have in the company. This approach also ensures that we do not lose the executives and employees—because if we were to start from scratch, they would rightly ask themselves whether the efforts of the last few months had not been worthwhile.

Once we have determined the current state with the analysis, we define the target state, which in turn can be determined from the corporate objectives. In a further sub-step, we determine the difference between the current state and the target state and derive a concrete need for action from this.

In concrete terms, this means the following sub-steps for us in the first phase:

- Determination of the actual state
- Definition of the target state
- Deriving the need for action

5.2.1.1 Internal Analysis

For the internal analysis, we first need to consider who we want and need to involve in the process. In my experience, two simple questions are useful for determining the group of people:

1. Who makes the decisions necessary to establish and implement the data strategy?
2. Who is particularly affected by the implementation of the data strategy?

Exemplary persons in the first group are company management, divisional management, or other employees who are responsible for data analysis. Once the group of people has been determined, the status quo must be established by means of an empirical survey among the groups mentioned. However, due to a cost–benefit ratio, no more than 50 people should be surveyed in total. The grid in Table 5.1 can be used as a guide for the survey. It can be seen that the skills and abilities are queried for both strategic and operational areas. A precise description of the group of people is not given here, as this depends to a large extent on the size of the company, its legal form, the sector and other parameters.

The selection of people in the second group is more difficult to determine, because ultimately many, if not almost all, employees will be affected by the data strategy in one form or another, completely independent of the task or hierarchical position. However, it is not appropriate to

Table 5.1 Questionnaire for internal analysis

#	Area	Criterion	Operationalization of the criterion
1	S4tatus quo, goals, and data strategy	Goals, strategy, alignment	Data & Analytics is imperative to the execution of our strategy and business model
			We have defined goals for the Data & Analytics area
			We have our own strategy for the Data & Analytics area
		Integration of and interaction with stakeholders	The relevant stakeholders for Data & Analytics have been identified
			The relevant stakeholders for Data & Analytics are involved in our projects
			Decision-makers are aware of the importance of data & analytics
			We know the needs of decision-makers for data & analytics
			The decision-makers in the company support Data & Analytics with appropriate resources
		Innovation orientation in the area of data & analytics	We have an innovation roadmap in the area of Data & Analytics
			We measure the effort and benefits of our innovation activities in the area of Data & Analytics
2	Organizational model	Organizational model	We have a formal organizational model (structure)
			This structure is suitable for our current and planned goals
			If the current structure is not suitable: we know how to adjust the structure

(continued)

Table 5.1 (continued)

#	Area	Criterion	Operationalization of the criterion
3	Process model	Collect	We have all the data we need to make decisions
			We use the relevant sources for data collection
		Understand	Our data is prepared in such a way that fundamental decisions can be made on the basis of this data
		Decide	Decisions in our company are made on the basis of the data generated
			The demand side can easily access the findings of the analyses
			The requirements of the demand drivers are taken into account in our data & analytics process
			The insights gained through data & analytics flow into business processes
		Automate	Our data & analytics process is automated
		Execute	We have processes and tools in place to evaluate the benefits of Data & Analytics
			The instruments capture quantitative and qualitative benefits
			The instruments establish a link between effort and benefit
4	Example projects	Use of data in use cases	We have use cases for the application of data & analytics
5	Tools	Identification and selection of tools	We follow technological developments with regard to tools that are relevant to us
			We have defined procedures/processes for selecting tools
		Tools used	Uniform tools in the individual areas

(continued)

5 Process Model for Implementing the Data-Driven Organization

Table 5.1 (continued)

#	Area	Criterion	Operationalization of the criterion
6	Data culture	Culture supports Data & Analytics	In our company, there is a willingness to share data
			The assurance of the quality of the data is ensured
			There is respectful interaction between Data & Analytics and the business departments
			When making decisions, data is involved
7	Talent management and talent strategy	Recruitment of personnel	We currently have the required capabilities
			We know in the short term what skills we need
		Education and training of personnel	We know in the medium and long term what capabilities we need
			We have clear career planning and development programs for our Data & Analytics employees
8	Data governance	Data governance	We have a clear structure for data governance
			We have defined roles for data governance
			We have processes in place for data governance
			We have processes in place to ensure data quality
			We have the necessary staff to ensure data governance

Source: Own representation

include every individual in the analysis. Rather, exemplary employees should be determined, for example, those who are responsible for data collection or data analysis. Once the group of people has been determined, the status quo must be established by means of an empirical survey among the aforementioned groups. However, in view of the cost–benefit ratio, no more than 50 people should be surveyed in total. The grid in Table 5.2 can be used as a guide for the survey. It can be seen that the skills and abilities are queried for both strategic and operational areas.

Table 5.2 Results of the internal and external analysis

#	Area	Rating	Biggest challenges	Greatest need for action
1	Data strategy and data governance	X points	Challenge 1 Challenge 2	Need for action 1 Need for action 2
2	Structure and communication	X points	Challenge 3 Challenge 4	Need for action 3 Need for action 4
3	Data & Analytics Process	X points	Challenge 5	Need for action 5
4	Tools	X points	Challenge 6 Challenge 7	Need for action 6 Need for action 7
5	Capabilities and data culture	X points	Challenge 8	Need for action 8
Total		Points XX	Total maturity	

Source: Own representation

For each of the operationalized criteria, respondents should be asked two questions:

- To what extent is the criterion fulfilled in our company?
- How strong should the criterion ideally be in the company?

By adding up the results for all respondents in each area and for all participants as a whole, statements can then be made about the need for action for each criterion. If, for example, the respondents rate the implementation of digital data archiving in the company on average as 3 (on a scale of 0–5) and they also see an ideal value of 3 for this criterion, there is no need for action. This is only the case if the ideal value is 4 or 5.

5.2.1.2 External Analysis

The external analysis does not have to be too comprehensive for the development of our data strategy. We can base our analysis on the points we have already looked at in Sect. 5.2 and seek answers to the following questions:

- Which technologies are relevant for our business?
- How can we use these technologies?

- What new business models threaten our business?
- Which old or new competitors pose a threat to our business?
- What are the best practices for data use in our industry? What can we learn from them?
- What does this mean for us overall?

5.2.1.3 Summary

Table 5.2 summarizes the results of the external and internal analysis. It represents one of the foundations for the elaboration of the objectives that will take place in the next step.

The table identifies the greatest challenges per area of analysis. These result from the difference (=gap) between the two answers of the participants (To what extent is criterion A fulfilled? How strong should A be in our company?). In a further step, the need for action can be derived from the challenges. Furthermore, an evaluation is possible for all participants as well as for individual participant groups (e.g., company management, division managers, digital specialists).

5.2.2 Data Targets

After the analysis we can define the goals. The starting points for defining the goals are the corporate goals and the results of the analysis. The corporate goals are important because the data goals and the data strategy derived from them later should not hang in a "vacuum." Rather, the corporate vision and goals should provide the anchor point for the data goals. But how exactly do we get from the vision and corporate goals to our own goals? We ultimately need to ask ourselves, for each and every goal, "How can I use data to support the business goals?"

To find your own goals, the following goal categories can help you:

- Optimization of processes (faster or better and thus in any case more cost-effective)
- Improvement of existing services to the customer

- New services for the customer
- Generate more sales
- Identify new business models
- Find new use cases.

A first starting point for the formulation of data goals are the corporate goals.

> **Example: Deriving Goals from Corporate Objectives**
>
> The automobile manufacturer BMW, for example, formulates one of its corporate goals as follows: "We work hand in hand internally and with our external partners." From this, we could directly derive several data goals that relate, on the one hand, to ensuring internal data exchange (e.g., reducing data silos) and, on the other hand, to supporting external collaboration (e.g., through appropriate cloud technologies). Alternatively, a data goal could also be to identify a certain number of use cases for data & analytics.

The results of the analysis represent a second starting point for formulating data goals. The goals can be derived from the identified challenges and even more concretely from the need for action of the internal and external analysis.

> **Example: Deriving Goals from the Analysis—1**
>
> One challenge identified in the internal analysis is that decision-makers in the company are not sufficiently aware of the importance of data and analytics, as the corresponding item ("Decision-makers are aware of the importance of data and analytics" from the area of "Data strategy and data governance") was only awarded two points, while the ideal score was seen by the respondents as five points. In this respect, a data goal should relate to increasing the commitment of the relevant stakeholders. Alternatively, a data goal could also be to identify a certain number of use cases for data and analytics.

> **Example: Deriving Goals from the Analysis—2**
> The analysis revealed that there are many data silos in the company and in the individual areas. A corresponding goal should therefore be to identify the causes of this and to create a common marketplace for the generated insights.

To check the formulation of data goals, the well-known and widely used SMART logic can be used. This sets the following five conditions for goals:

1. Are the objectives specific and well understood by all stakeholders and affected parties? (Specific)
2. Are criteria definable that allow verification of goal achievement? (Measurable)
3. Are the goals challenging enough to motivate employees? (Achievable)
4. Are the goals realistically achievable? (Realistic)
5. Are time points defined by which the individual goals must be achieved? (Time-bound)

Once the goals have been formulated, the next step is to condense them. In many cases, the analysis alone results in a large number of goals. The following procedures are suitable for condensation:

- "Marrying" two or more substantively similar goals into a single goal.
- Prioritize goals by developing a ranking and focusing on the most important goals.

As a result, we obtain a manageable number of targets (approx. five to ten targets) that must be confirmed by the company's management. At the same time, this results in a mandate to develop a data strategy.

5.2.3 Data Strategy

5.2.3.1 Definition

A strategy basically shows a way how goals are to be achieved. In this way, the strategy forms the bridge between the goals, on the one hand, and the concrete actions on the other. Strategies always make statements on specific areas: a corporate strategy, for example, on the service portfolio, a business unit strategy on the behavior vis-à-vis the competition. And a data strategy also makes statements on a number of data-related topics, which I would like to introduce to you in the following. For example, the make-or-buy decision within the data strategy provides a basic direction for the individual areas of the company in order to simplify decision-making. The data strategy thus provides a corridor within which downstream decisions should move. Without a data strategy, an extensive evaluation process would have to be carried out for each individual data-related decision, and there would be a great risk that the individual decisions would not be consistent.

> **Data Strategy**
> A data strategy is a plan that identifies how we, as an organization, will use data as an asset to achieve our business goals.

> **Example: Data strategy**
> One area of the company decides at the operational level to purchase an expensive software solution, while another area decides to develop the solution itself. In sum, this results in a higher effort compared to the purchase of a somewhat more extensive and expensive solution, which, however, fulfills the requirements of both areas.

5.2.3.2 Delimitation

The data strategy must be clearly distinguished from the company's digital strategy: The latter is hierarchically higher and describes how the company wants to manage the digital transformation. Typical statements of the digital strategy relate, for example, to the use of new channels for communication and interaction with customers.

5.2.3.3 Development of the Data Strategy

For the development of the data strategy, which is derived from the corporate strategy, we have to ask ourselves two questions:

1. According to which mode should the data strategy be developed?
2. Who should collaborate on the strategy?

Three different modes are available for developing the data strategy with regard to the first question.

First, we can plan the data strategy top-down. This means that the strategy is developed by the higher level and subsequently concretized for the subordinate levels. The advantage of this variant is that there is a high degree of consistency in the sub-plans for the individual areas, and there is also a speed advantage. However, there is a risk that the superordinate level will assume unrealistic targets due to a lack of knowledge of the subordinate areas.

The second mode, the bottom-up approach, chooses the opposite path: planning starts at the lowest level and is passed on to the next higher level, where control, coordination, and supplementation take place. With this variant, employee motivation increases due to the involvement of the levels. Furthermore, the framework conditions of the individual levels can be taken into account. This increases the feasibility of planning compared to the first variant. Disadvantages are:

- Contradictions between the individual plans as well as
- a large amount of time required for coordination and integration of the sub-plans.

Which of the two variants is the "better" one cannot be judged per se. However, it is also possible to combine the two approaches (=countercurrent method). In this case, the higher level provides a rough framework (top-down approach), this is elaborated in more detail by the subordinate areas and subsequently passed upwards again for coordination and integration (=bottom-up approach).

The second aspect, i.e., the question of the people to be integrated, must be considered depending on the mode. If a top-down approach is chosen, only a small circle is responsible for strategy development. However, it can also be considered here whether subordinate levels should also be involved in accordance with the procedure already described in the analysis ("Who makes the decisions?" and "Who is affected by the decisions?"). It makes sense, for example, to integrate middle management. Here it often helps to draw up an organization chart and circle those people who will work with or provide the data that will later be relevant, and to involve these people in the process. We will need the organization chart with the marked persons and their tasks later for further steps.

5.2.3.4 Statement Areas of Strategies

Make or Buy

A first statement area concerns make-or-buy decisions. A make-or-buy decision is about whether a company creates services itself or buys them externally. In the operational area, the make-or-buy decision is particularly relevant for tools and instruments as well as personnel. In the case of tools and instruments, the question is whether solutions should be created using in-house resources or whether ready-made solutions should be purchased. With regard to personnel, the question is whether talent should be hired or recruited and trained from within the company. Strategic buy decisions are primarily aimed at concentrating on the company's own core business and outsourcing non-core activities. Make-or-buy decisions go hand in hand with the development of a company's own competencies and independence from the market. Whether the make-or-buy strategy is more favorable for a company depends on the general conditions and the maturity of the company. For example, in the case of a small company with only a few employees, it may make more sense to buy a standard solution than to attempt to program it yourself.

> **Example: Make-or-Buy Decision**
>
> An example of a make-or-buy decision at the strategic level is the question of whether a company connects its data sources itself and thus accepts the development effort and the expense of support, or whether the company buys a standard tool on the market for this purpose.

Customer Access

A second statement area of the data strategy concerns customer access. A company basically has two options: Either the company has direct customer access, i.e., a customer interface through which direct communication, interaction, and also sales of services are possible, or the company purchases access to the customer from other companies. Having one's own access is generally associated with a higher level of effort (e.g., gaining and retaining customers, regular interaction, provision of suitable solutions …), but in return there is a high degree of independence. If customer access is acquired, a consideration must usually be paid for it (e.g., commission on purchases), and there is also a dependency on the actor who has customer access.

> **Example: Make or Buy Online Shop**
>
> A retailer can operate his business with both strategy types. In this example, a separate customer access would be a separate store with an additional online store. The retailer promotes the online store via various channels, such as social media, search engines, etc. The retailer's own customer access is also used. A retailer without customer access, on the other hand, uses third-party platforms such as Amazon Market Places, the Real marketplace, or eBay.

The advantage of direct customer access is that the company can better understand the market and respond much more quickly to customer needs. The company therefore has its "ear to the customer" and is much more aware of developments and trends. One disadvantage is certainly that setting up and maintaining the structures, systems, and processes to ensure direct customer contact involves a great deal of effort. For example, a company with customer access has higher expenses for personnel, for marketing, for updating or expanding the systems and, last but not least, for data protection and data security.

5.2.3.5 Success Factors for the Development of the Data Strategy

I see a number of success factors for the development and implementation of a data strategy. An old hat, but nevertheless the most important point, is the involvement of top management. Without the top decision-makers, no strategy, not even a data strategy, can get off the ground. Secondly, the strategy should be future-oriented as well as feasible and value-creating. Another success factor is that we break down data silos. This is not an easy process, as we first have to identify the responsible people and bring them to the table. It has also turned out to be important to proceed with a "salami tactic," i.e., to implement simple and quickly realizable projects in order to then be able to sell the added value of data & analytics with simple data stories. Finally, one should proceed according to the Pareto principle for many steps. In this context, this means: 80% planning is perfectly sufficient, we don't want to "die in beauty." It is more important to get into action.

You can use the following list as a checklist for your data strategy.

> **Data Strategy Checklist (Based on U.S. Department of Defense, 2020, p. 2)**
>
> - **Visible**: The users can quickly locate the required data, i.e. they can quickly find the required data.
> - **Accessible**: The users can easily and quickly access the data they need.
> - **Understandable**: Descriptions are provided to stakeholders to help identify and understand the data.
> - **Linked**: Requirement owners can identify interrelated data through appropriate cues.
> - **Trustworthy**: Demand drivers can rely on the data.
> - **Interoperable**: The demand side has a common and unified understanding of the data.
> - **Secure**: Data is protected from unauthorized use and access.

5.2.4 Stakeholder Integration

In every book on strategic planning, project management, or even publications on data & analytics, stakeholder engagement is considered one of the most important tasks and often the basic prerequisite for success. But why are stakeholders so important? First of all, we have to ask ourselves who the important stakeholders are for us.

The challenge in connection with stakeholders is that they are often caught up in their current day-to-day business and have neither the time nor the leisure to think outside the box. So, by behaving this way, they prevent new technologies and/or approaches from finding better solutions to their current problems. Other challenges are that stakeholders often pursue their own agenda and/or have respect, fear, or even a lack of trust to try something new.

5.3 Organization Model

The structure of a company, also referred to as the organizational structure, results from a break down of the overall task of a company into individual subtasks and an assignment of these tasks to a job. These jobs are combined to form an overall structure, which can then be depicted in an organizational chart. This means that the structure of a company does not exist as an end in itself, but ideally represents the best possible structure for implementing the company's task. However, this also means that the structure cannot be regarded as fixed, but can or must change depending on the goals and strategy of a company ("structure follows strategy").

To implement our data strategy, we also need a structure as a first step. More precisely, we need to determine how our core team should be organized and work together with the "rest" of the company. Three alternatives exist for this.

1. The first option is centralized implementation: The data and analytics experts form a team located in a single place in the company and take on advisory tasks for the business units. This approach can be implemented quickly and also more cost-effectively than the other

variants. As a result, different types of analysis (especially statistical analyses, data mining models, predictive analytics) can be deployed quickly. The disadvantage, however, is that priority and resource problems can arise because the analysts cannot support all areas at the same time. Furthermore, there is a lack of direct contact with the individual divisions, which means that analysts may not be able to understand the problems and wishes of the divisions to a sufficient extent. Finally, there is also the question of which area the data & analytics unit should report to (management, marketing, finance). This approach is particularly suitable for companies that are new to the field of data and analytics, as it can be used to achieve initial success very quickly.
2. From a structural point of view, the embedded analyst approach is exactly the opposite of the centralized approach. Embedded analysts are teams of analysts who are permanently assigned to individual business units. The advantage of this approach is that there is no wrangling over priorities or resources due to the fixed allocation. Prioritization is only necessary between different data and analytics projects within a division. Due to the proximity to the individual business units, the decentralized approach offers the advantage that the analysts can get a very good picture of the tasks and challenges of the business units. However, one challenge of this structure is to gather the necessary manpower to tackle and implement company-wide tasks and projects. A further disadvantage of this approach is that analysts often adopt a departmental perspective and lose sight of the big picture. This can and must be counteracted with a regular exchange with other areas.
3. The third option is known as the hub-and-spoke approach. On the one hand, this approach comprises a central team, which assumes coordinating tasks, for example, and, on the other, teams of analysts in the individual business units. This approach thus combines the advantages of the two previously mentioned approaches, but also involves risks:

Table 5.3 Comparison of the organizational models

	Potentials	Risks
Centralized approach	Quick and inexpensive to implement Infrastructure easy to set up	Setting priorities/allocating resources to supported areas Lack of understanding of the business units due to distance
Decentralized approach	Clear prioritization and allocation of resources Good understanding of the business units	Implementation of company-wide tasks and projects
Hub-and-spoke approach	Good understanding of the business units Priorities and resource allocation established Implementation of company-wide tasks and projects	Coordination between central and decentralized units?

Source: Own representation

- Coordination between central and decentralized units.
- Exchange between the individual analysis teams is not optimal.
- Spokes no longer see responsibility in themselves, but in the central team.

Table 5.3 summarizes the potentials and risks of the three approaches.

In my experience, the hub-and-spoke approach has proven to be very effective because of the advantages of the hybrid model.

Regardless of the approach chosen, I believe it is essential to understand data & analytics as a tool that can and should support the solution of problems and create value for the company. It is therefore important that Data & Analytics does not work in isolation from the business departments and sits in an "ivory tower," but develops solutions and productive tools that are accepted by the users because of the added value they generate. Data & Analytics must therefore see itself as an enabler for the business departments.

> **Example: Structuring the Collaboration Between Data & Analytics and the Specialist Departments at an Apparel Manufacturer**
>
> The task of analysts is to provide data with which the specialist departments can do their work better. Data & Analytics ensures that the methods used to collect data meet quality criteria and that the data is cleaned and prepared. However, Data & Analytics is not the "police" who check the work of the specialist departments and try to find errors or uncover misconduct. Nor is it the task of the data area to patronize the specialists or even to make decisions itself. Data & Analytics should, however, provide data that enables better decisions. However, the prerequisite for this is that the specialist departments trust the data and do not question data, for example, simply because it contradicts their own opinion or world view. Ultimately, for many companies, future competitiveness depends on the collection, analysis, and use of data from many internal and external sources—but this use can only take place as part of a collaborative project between data science and specialist departments (see Rashedi & Feng, 2021).

5.4 Process Model

> **To ensure good collaboration and to ensure added value, Data & Analytics should be**
>
> - ... interact with the specialist departments and network within the company. And not just sit in the ivory tower and try to develop solutions on their own that don't meet the actual needs of the departments.
> - ... see themselves as enablers for the business departments and do not have the claim attitude to understand the business better. Rather, Data & Analytics is the consultant that enables the business departments to use the data and makes itself available as a sparring partner.
> - ... have the claim to themselves not to analyze data, but to change the company. To do this, Data & Analytics must analyze the needs of the departments, make suitable offers, and promote its own solutions.
> - ... not ask the question of what can basically be done with the (existing) data, but answer the question of where the company wants to go, where the greatest inefficiencies lie, what new business areas exist and how data can support with all these questions.
> - ... do not necessarily apply the latest and most complicated methods, but proceed pragmatically and develop solutions that bring the required result. In concrete terms, even simple static but at the same time clever algorithms can already deliver a high added value.

5.5 Example Projects

5.5.1 Self-Services and Real-Time Services at the Schwarz Group[1]

In a first example project, Dimitri Malinov talks about self-services[2] and real-time services. He notes an increasing interest in self-services, so that the topic is becoming more and more relevant even compared to the current hype in the area of data and analytics, AI. He attributes the increasing relevance to the following reasons:

- **Easy data availability**: Meanwhile, all processes (internal to the company, external to the company with reference to customers or suppliers) generate digital data. As a result, all processes are traceable and transparent via the associated data flows (e.g., customer journey in the online store).
- **Experience in dealing with data**: In companies, there are more and more employees who, for various reasons, have experience with data and for whom handling data is no longer an exception.
- **Rethinking in companies**: Compared to the past, silo thinking in companies has decreased and individual divisions/departments are more willing to share their data. The divisions/departments have recognized that data is an enabler for cross-divisional collaboration. The prerequisite is that there has been agreement on the data to be used.

With regard to the implementation of self-services, Dimitri explains that the interest often comes from the departments themselves ("I think the reporting is good in principle, but I would like to have the statements from this Excel file to supplement it."). In his opinion, a project can be started with this interested group of people from the departments ("friendly users").

As a goal-oriented approach to implementation, not only should the tools be made available to the departments, but the employees must also

---

[1] Cf. Rashedi and Malinov (2021).
[2] Self-services in this context means that users can independently access the data relevant to them and independently create both analyses and reports.

be empowered to use the tools. According to Dimitri, this achieves three positive effects:

- For employees, the hurdle to using the tools is lowered, as employees are often just afraid of contact.
- By using their own tools for the first time, employees can create their own analyses or reports that mean added value for them. In the best case, these own solutions are communicated to others ("Look what I created myself"), so that employees become ambassadors for self-service tools.
- Business departments often don't initially know exactly which reports or analytics they really need and what will add the most value for them. By trying things out on their own, they can experiment and see what would be helpful for them. They can already implement some of the analyses themselves using self-service tools; for the rest, they can ask Data Science for support with clear objectives.

In order to convince management of the benefits of a self-service, a quantitative analysis can be carried out. To do this, the time required for the manual creation of reports for the area under consideration must be recorded and extrapolated per month. This expenditure can be compared with the resources required to implement a self-service project. As a rule, the implementation of a self-service pays off after 2–3 years.

For the concrete structuring of self-service reporting, the following procedure is recommended:

- Start of implementation in an area with "friendly users"
- Development of a limited number of templates for all areas as a basic structure (e.g., six to ten templates)
- Implementation of the developed basic structure of reporting by software
- Training of employees on this software
- Filling of the basic structure by the employees of the selected area
- Communicate reporting throughout the organization to "hook" people and build momentum
- Roll out self-service throughout the organization.

Standardization across all areas means that reporting has a uniform basic structure and can be understood more easily. The following example shows the added value created by self-services.

> **Example: Added Value of Self-Service**
>
> By using self-service tools, departments can get the information they need much faster. Let us assume, for example, that for an online store the availability of articles is to be checked and the reasons for non-availability identified: A single department, such as sales or logistics, cannot answer these questions because only their own data is available in each case. Through cooperation between the areas and the use of self-service tools, the necessary data can be generated much more quickly.

With regard to real-time services, Dimitri argues that they are needed in two situations in particular: First, these services can be applied to processes that run very often (e.g., repetitive processes in manufacturing). In these processes, real-time data helps in troubleshooting and making decisions: Previously, data was only available in analog form with a significant delay; today, data can be viewed in real time on the screen and used as a basis for decision-making. However, the use of real-time services is not only useful when decisions are to be made immediately on the basis of the data. Rather, secondly, a user may simply want to have up-to-date information, even if the decisions are not made immediately (e.g., monitoring the launch of a new website).

The following example shows a typical application area for real-time services in connection with an online store.

> **Example: Typical Application area of Real-Time Services**
>
> An exemplary area of application for real-time services is online stores, as automatic segmentation can be used to create a homepage tailored to the customer (=liquid homepage). This can lead to an improved customer experience, as the customer is shown content and offers that are suitable for him. Ultimately, the combination of AI and real-time technology creates a situation in which the customer is confronted with his longstanding salesperson from the offline world, who knows his preferences precisely and can assess the customer.

5.5.2 Data & Analytics in B2B[3]

Romy Beer comments on the implementation of data & analytics in a B2B company. She argues that B2B companies lag significantly behind their B2C counterparts in the data area. Some of the innovations, processes, and tools used in the B2C area only arrive in the B2B area with a considerable delay. For example, the omnichannel approach to ensuring a consistent brand experience across all available channels has been standard in the B2C sector for years; in the B2B sector, only a few touchpoints are often integrated.

However, the users "function" in a similar way to those in the B2C sector; ultimately, they are the same people who merely act in different contexts. In this respect, engagement in social media can also lead to success in the B2B sector. According to Romy, the main differences between the B2B and B2C sectors with regard to data & analytics are the following:

- Fewer visitors to the web presences and thus also less data
- Targets are mostly leads and not conversions
- Use of own platforms in the B2B area
- Opportunity to experiment, as the number pressure is not yet so high and the change is slower
- Less intensive use of tools
- Lower customer engagement.

According to Romy, the starting point for setting up data and analytics in the B2B sector is often a situation in which sales are still running completely offline. In this situation, the basic infrastructure must first be built ("paving the road to business"). After that, initial attempts can be made on marketplaces such as eBay or Amazon, as interested parties for a B2B company's products can also be found there. However, since these platforms do not provide data, the next step is to establish a company's own presence in the form of an online store. The company's own data can then be generated there and the evaluation can begin in a step-by-step procedure.

---

[3] Cf. Rashedi and Beer (2021).

5.5.3 Configuration of the Collaboration Between Data & Analytics and the Business Departments at a Fashion Company[4]

On a possible configuration of the collaboration between Data & Analytics and the business departments at a fashion company, Christopher provides insights.

The fashion company has set up a data science program aimed at developing various analytical models. Structurally, the program manifests itself in several virtual teams.

These teams include data analysts as well as business analysts and people from the specialist departments. Although the people from the data area do not necessarily know every detail of the tasks and problem areas of the business departments, they can translate a concrete case from the business departments into use cases. The advantage of this approach is that data and business analysts have a greater distance from the business and can therefore often identify problems faster or better. Data analysts continue to provide the infrastructure. The teams work collaboratively on this infrastructure and also make the results available on it.

In terms of content, the virtual teams deal with different subject areas and work on solutions related to, for example, demand forecasting processes, recognizing trends, identifying the causes of delivery problems, or finding optimal solutions for the design of the supply chain.

The fashion company applies the following criteria for the selection of use cases:

- **Dependencies in your own value creation process**: There are many dependencies in the value-creation chain of every company. These dependencies must be understood, since a use case can only be meaningfully processed if the upstream issues in the value chain are resolved.
- **Significance of the use case**: Some use cases are very important for a company because, for example, they occur in several areas of the planning process or are relevant for several different functional areas (e.g., determining price elasticities).

---

[4] Cf. Rashedi and Barth (2021).

- **Relationship between effort and return**: Controlling can be used to answer the question of what positive financial and non-financial impact the solution to a use case has. This impact can be compared with the estimated effort.
- Last but not least: **Assessment of Data Science.**

To build Data & Analytics, the following steps were taken in the fashion company:

- Concrete need for action identified: Business unit cannot continue to work with the existing process (e.g., because the scope of data has changed or new strategic challenges have arisen).
- Focus on a limited number of particularly important use cases.
- Detailed elaboration of these use cases.
- Internal marketing for the developed use cases. The objective is for employees to gain confidence in the data and the insights generated from it.

5.5.4 Re-launching Data & Analytics at a Content Provider in the Sports Sector

The next example project[5] focuses on a sports platform. The platform used Google Analytics exclusively. However, there were numerous challenges (e.g., double counting in the products, incorrect assignment of visitor sources). The generated data was transferred to Excel to create presentations for management. However, due to the challenges cited, it was not possible to make control statements (e.g., which part of the marketing budget should be used for customer acquisition via which channels? Which user segment contributes to the company's success and how?)

Data & Analytics therefore sets itself the goal of collecting the data cleanly and transferring it to a data lake in order to subsequently be able to make statements on control and to make predictions.

---

[5] Cf. Rashedi and Damm (2021).

In a first step, the tool was changed (ATI) and operated in a cloud. Since the "old" data was no longer used due to the uncertainties it contained, only a few statements could initially be generated. Subsequently, all value-creation levels of the company (editorial, sales, product management, finance …) were mapped and self-service was enabled via Power-BI. This made it possible to establish a uniform understanding within the company of the interrelationships in the value chain.

It was also possible to identify different user segments for which specific statements could be made about behavior (e.g., share of the segment in total users, type of products used, such as livestreams or reading articles) and about the revenues generated with the respective segment. On the basis of this information, it was possible in a next step to provide indications for the development of new products, in particular for the development of pay products, with the inclusion of further external data. Data & Analytics was able to clearly show what revenue potential exists compared to simply playing out advertising. This represented a fundamentally new approach for the company, as it meant switching from an advertising-financed model with a focus on reach to a model based on customer centricity and sales products.

This procedure was the prerequisite for determining the user lifetime value. The user lifetime value currently refers to advertising revenues. For this purpose, data from various sources are combined:

- Analytics data
- Marketer data
- Article data
- Online marketing data
- Usage behavior (types of sports, content used).

Regression analyses can then be used to make statements about the user lifetime value.

5.6 Tools

The tools to be used for Data & Analytics depend on the specific requirements for the data area. Numerous vendors have now established themselves on the market, offering very specific tools for the individual areas of the data lifecycle. The following is not intended to promote specific tools from individual vendors. Rather, I would like to present basic categories for which data & analytics tools should be available. These criteria are (cf. Goyal, 2021):

- **Behavioral data ingestion**: These are tools that are able to capture customer behavior along the customer journey. This enables an understanding of user behavior to be built up.
- **Transactional Data Ingestion**: Tools to bring together data from disparate sources to build a 360° understanding of the customer.
- **Storage**: Scalable data storage solutions (cloud data warehouse and cloud data lake) that enable low-latency data access.
- **Processing**: Solutions that enable Data & Analytics to aggregate and filter raw data sets in the various storage media in order to perform analyses.
- **Operations**: Reverse ETL solutions that enable further processing of data from the data warehouse in different solutions (e.g., CRM solutions) and support self-service.
- **Analysis**: Solutions that support data analysis through various visualization and exploration functions.
- **Intelligence**: Solutions that support the creation and testing of models that can identify relationships in historical data and predict future developments.
- **Management**: Solutions to manage data pipelines, monitor data quality, and address organizational challenges.

A number of tools from these categories can only be used by experts. However, when selecting tools, I think it should be checked whether easier-to-use alternatives exist. To illustrate this with an example: In the meantime, there are tools available under the name Auto-Machine

Learning (AutoML) that enable users without a decided knowledge of programming and software to use machine learning. Specifically, these tools can be used to automate either individual process steps of machine learning (e.g., selection of algorithms suitable for the task) or the entire machine learning process. The use of AutoML tools can reduce the need for data specialists (cf. Luber & Litzel, 2020).

The following example provides an overview of tools for use in the social media space:

> **Example: Tools in the Social Media Area at a Clothing Company**
> The prerequisite for the analysis are tools for the different data sources (e.g., Facebook, Instagram, Twitter …). These tools must necessarily be used, since rules exist for the handling of data and own scraping of the data is not allowed. In this respect, different third-party tools are used, e.g.:
>
> - PR Scraping Tools
> - Tools for analyzing users on your own pages to analyze visitor behavior
> - Tools to manage social media channels and measure performance (e.g., reach, engagement …)
> - Experimental tools, e.g., for Natural Language Processing or for analyzing image content
> - SM channels to manage activities and analyze performance (e.g., reach, engagement) (cf. Rashedi & Feng, 2021).

5.7 Data Culture

In the past, the relevant specialist literature and research on the topic of BI dealt predominantly with technological issues. However, due to challenges in the implementation of BI in companies, the human factors and corporate and data culture are increasingly gaining attention (see Talocka et al., 2018). A similar picture emerges in practice. For example, a study from 2019 shows that companies are pursuing the objectives of implementing data-driven corporate cultures and anchoring the topic of culture at both the strategic and operational levels. In the analysis cited, the topic of corporate culture advanced to become an important trend; among North American companies, the establishment of a data-driven corporate culture even has the highest significance (see BARC, 2019, pp. 11 and 19).

Talocka et al. (2018) differentiate three different types of cultures in the context of BI:

- The corporate culture, which includes values as well as norms, behavior, and habits in the organization.
- The information culture is considered to be part of the organizational culture. It includes all processes in the company related to the processing of information, as well as those values, norms, behavior, and habits that exist in the company with regard to the processing, use, and sharing of information.
- The BI culture is part of the information culture. It includes the values, norms, behavior, and habits that promote (catalyze) the provision of decision-relevant knowledge to users.

Characteristic features of a BI culture are:

- **Information sharing culture**: Information sharing refers, on the one hand, to the people involved in the BI process, but also to the sharing of data and mutual support in the use of tools. Furthermore, information exchange means opening up the data silos that exist in the organization.
- **Creation of a business intelligence community**: Both models and the results of BI processes should be shared in this community.
- **Sharing positive and negative experiences**: This can be done, for example, by establishing a company-wide platform that is accessible to all and on which success stories, but also mistakes and failures, are documented so that the experience already gained is not lost to the company.

Various approaches and models exist for capturing corporate culture. A very clear and pragmatic model is the water lily or culture level model (see Fig. 5.2). The model goes back to the U.S. organizational

5 Process Model for Implementing the Data-Driven Organization

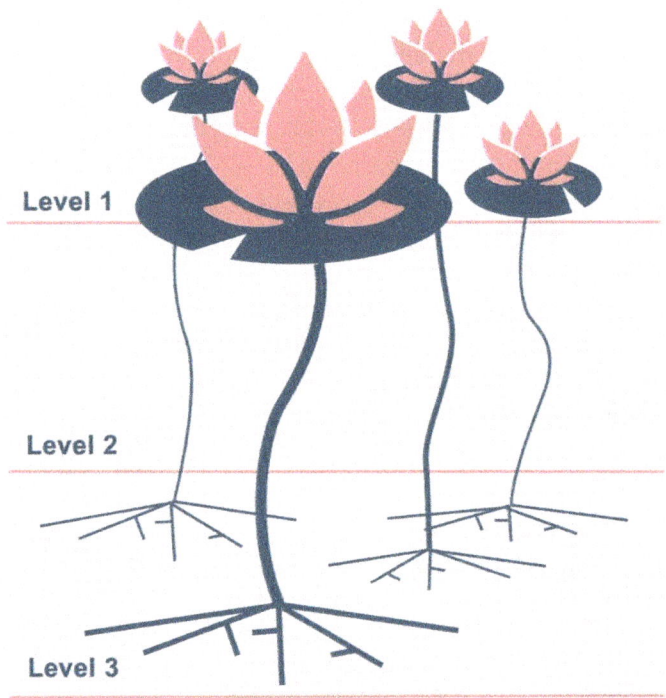

Fig. 5.2 Culture level model (Source: Own presentation based on Stolpe & Hebing, 2020)

psychologist Schein (2016, pp. 17–30). However, while Schein applies the model to describe corporate or organizational culture, it will be used below to characterize BI culture. As can be seen in Fig. 5.2, the model differentiates between three levels.

The first level refers to observable behaviors as well as the physical manifestations of BI culture, e.g., in the form of artifacts or products. These can be printed reports or information dashboards, for example. Another manifestation may be individuals at the executive level who insist that decisions made by subordinate areas be data-informed. Schein compares this level to the leaves and flowers of a water lily, which are visible to everyone on the surface of the pond. The second level includes the stems of the water lily, which lie below the surface of the water and can sometimes be seen by an observer, given favorable refraction of the light. Alternatively, a person would have to dive into the lake to see the stems. In a figurative sense, the stems are responsible for providing a link between the first and third levels. Projecting this onto the BI context, this level manifests itself in the values and norms of BI. Specifically, these can be instructions for action or ethical principles related to data governance. The third level, ultimately the roots of the water lilies, can be equated with the existing basic assumptions and imprints that influence thinking and action in the company in the BI context without being consciously perceived by the employees. An example of this could be the (BI-hostile) basic belief of employees that sharing data endangers their own position in the company. A BI-friendly imprint might be "If you give me your data, I'll give you my data and we'll both fare better than before."

> **Example: Data culture at a Clothing Manufacturer from the Sports Sector**
>
> The best possible collaboration between Data Science and business departments is based on mutual appreciation and trust. This starts with the type of request: The business department can ask Data Science to "build a dashboard for XY" and Data Science delivers. However, it is more effective if the business department explains its own goals and the context. Then Data Science can show what data is available and how the business department can be supported. In this way, Data Science does not act like a "call center" that sends data on demand, and the department has a better feeling because the explanation of goals and the context ensures that the data and insights are delivered (see Rashedi & Feng, 2021).

The challenge in connection with a corporate culture is now that it eludes direct control: Unlike a vision, a goal, or a mission, a culture is always something that has grown and not something that is defined. Nevertheless, a corporate culture and a data culture can be acted upon. The model described can help analyze the culture that exists in a company and derive recommendations for action to shape it. In the context of creating a data-driven organization, the levels described could be addressed as follows:

- Installation of real-time dashboards in the company, through which every employee can read the development of important performance indicators (= first level, dashboard represents an artifact).
- The dashboards convey to employees that data is not a secret in the company and that every employee is granted insight into key developments (= second level, sharing data as one of the core corporate values).
- Implicit in this approach is the long-term attitude that every employee, regardless of hierarchical level, has access to the numbers they need to make data-informed decisions.

I also use this procedure and can give a hint for the practical implementation of the four phases. The first point (planning) should end with the identification of people or departments that already have experience with data and where the creation of the user experience can be realized quickly in the second step (quick wins, low hanging fruit).

5.8 Talent Management and Talent Strategy

In order to implement a data-driven organization, suitable employees are indispensable. In this context, the following questions arise:

1. Which employees with which qualifications do I need?
2. How many employees do I need?
3. Where do I get these employees?

Let us start with the first question: Four different qualifications are essentially required for the implementation of data & analytics. The Data Engineer ("data engineer") has technical as well as mathematical-statistical and IT skills. He or she is responsible for identifying the data sources required to answer the questions posed, merging and cleansing the data obtained, as well as performing analyses and communicating the results to those who need them. The Data Scientist has a similar scope of duties as the Data Engineer, but focuses on advanced analytical methods in the sense of advanced analytics (e.g., prescriptive and predictive analyses). The data analyst ("data analyst"), as the third function in the data team, also performs analyses, but this person does not work with raw data but with data that has already been prepared. The business analyst/ops analyst represents the interface between the data area and the business departments. His area of responsibility is the improvement of processes and systems as well as the answering of specific questions from the business departments or the creation of dashboards. Often, the Business Analyst/Ops Analyst works with a specific functional area (e.g., Finance or Marketing). The Head of Data Analytics integrates the functions described above and is responsible for the strategic development of data expertise within the company. He or she also ensures data governance and is the point of contact for corporate management (see de Leyritz, 2021).

The specific qualifications required in the data team depend on the type of request. If there is no dedicated data & analytics department in the company beforehand and this is to be established first, a data engineer and a data analyst are often sufficient. The same applies if the requests relate exclusively to the analysis of past developments and the understanding of these developments. Regardless of the nature of the requests, however, the team should have skills in databases, software development, machine learning, and visualization. Furthermore, it may be necessary to integrate data specialists with industry knowledge or specific skills into the team (e.g., for the analyses related to an online store or the company's value chain).

In my experience, it has also proven successful to rely on people who can "do" and drive the first projects when setting up Data & Analytics. This ability is initially more important than creating elaborate analyses. Subsequently, it becomes more important to have people in the team who have a well-founded data background according to the above categorization. In addition, there should be a focus on individual use cases that can subsequently be continuously expanded.

The "right" size of the data team depends primarily on the size of the company, the maturity of data & analytics, and the number of ongoing and planned projects. The size of the company has an influence in that, as the size of the company increases, not only does the number of requests to the data area increase, but also the complexity of the questions to be answered. Also, as maturity increases, so do the analytics requirements, which also argues for a larger data team. However, concrete deductions for team size are most likely to be made based on current and planned projects: If you locate the projects on a timeline, map them with the available resources, and determine that they will take longer than 9 or 12 months to complete, then you should think about increasing the size of the team.

The third question to be answered relates to the acquisition of employees. This is a make-or-buy decision in that a company can either train the employees itself or buy trained personnel on the labor market. The challenge in acquiring staff on the labor market is that the number of job postings for Data Scientists has roughly tripled since 2013. For established companies in particular, it is difficult to compete with the large technology companies and secure the employees they need, according to McKinsey, citing an analysis of indeed.com (job search engine) (see Hürtgen et al., 2020). One possible strategy for recruiting staff can be to enter into partnerships with colleges and universities in order to recruit working students for smaller projects and then, in the best case, to recruit these people for the company after they have completed their studies. For the training of employees in the company, a mix of organized continuing education (internal continuing education, attendance at external continuing education courses) and employees' own continuing education (e.g., online continuing education via platforms such as Udemy) is a good idea.

> **Competencies for Work in the Data Area**
>
> - **Technical know-how**: The more pointedly the area is positioned, the more important is the technical know-how.
> - **Communication**: The broader the field, the more important communication becomes. Communication also includes the ability to translate the language of the analysts into the language of the business units. This refers, for example, to the translation of English-language metrics into terms that are familiar to users and whose content they can grasp immediately.
> - **Analytics**: Ability to perform analytics in respective areas (e.g., web analyst).

Indispensable for data & analytics is also a long-term talent management with a clean process and a clearly defined strategy. Figure 5.3 shows a possible process for talent management that can also be applied to the data area. The first step of planning is based on the corporate strategy as well as the data strategy. The goal is to determine future needs based on the current workforce. This is to ensure that the data area has the necessary skills to implement the strategy. Based on the identified needs, the second step is to recruit the necessary employees, either from within the company or from the labor market. In the best case, the company succeeds in establishing a unique selling proposition (cf. Momtasian, n.d.).

In addition to onboarding, development (third step) primarily involves performance appraisals, the further development of employees along various paths, and the establishment of career paths. Career paths can be used to show employees what planned development they can undergo in the company. This also contributes to employee retention. Other measures to prevent churn include an appropriate corporate culture and a

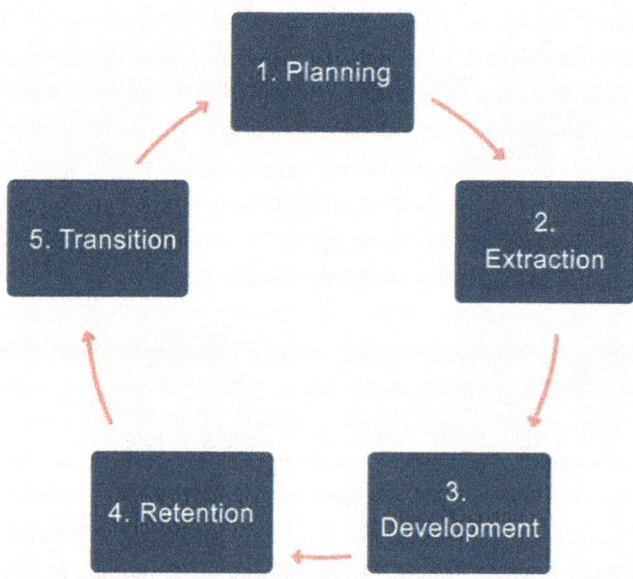

Fig. 5.3 Process for talent management (Source: Own representation based on Momtasian, n.d.)

Table 5.4 Components of a talent strategy

#	Area	Measures
1	Employer branding	Build a strong employer brand and implement recruitment campaigns that address the needs of talent or graduates
2	Quality management	Continuous internal analysis to survey employee experience and define actions to identify and develop talent
3	Involvement of managers	Evaluation of managers not only in terms of their professional performance, but also with regard to (pre-)life and the communication of core corporate principles and values
4	Enabling executives	Further development of managers (e.g., development of coaching skills of managers)
5	Talent evaluation	Implementation of a tool or process for the identification and assessment of talents
6	Internal networking	The realization of an internal networking and exchange of topics related to talent management
7	Continuous further development	Further development of own talent management through a continuous search for innovations

Source: Ready et al. (2014)

compensation strategy for employees. The fifth and final step involves succession planning, retirement planning, knowledge management, and exit interviews (see Momtasian, n.d.).

The process outlined is ultimately a pure process model that can be used in any company for different functional areas. Talent management only becomes specific through an individual talent strategy of the company, which makes concrete statements on the long-term orientation and implementation of talent management. Table 5.4 shows tips for designing a talent strategy that have been developed on the basis of best-practice examples and can also be applied to the data area.

5.9 Data Governance

The final step in the process is to address the very important issue of data governance. At the latest when reports are questioned in the company or when metrics or KPIs appear in reports that are defined differently or interpreted differently by the users, then you should think about data governance.

The Data Governance Institute defines the term as follows:

> a system of decision rights and accountabilities for information-related processes, executed according to agreed-upon models which describe who can take what actions with what information, and when, under what circumstances, using what methods (Data Governance Institute, 2020).

Data governance thus describes a system of processes, specifications, and roles that regulates the management and use of data in the company in order to take sufficient account of both security-related and quality-related requirements. In particular, data governance regulates

- which persons or roles in the company
- for what purposes
- in compliance with which framework conditions and
- using which methods
- which data may be used.

Or to put it more simply: data governance provides a framework for handling data. For me personally, however, data governance is **not a** paper in the form of a formulated guideline.

Data governance is to be distinguished from:

- **Data stewardship**: Data stewardship refers to the implementation of data governance at an operational level. In other words, it helps to ensure data quality and the quality of data sources, and is responsible for the ease with which data can be found and processed.
- **Data management**: The aim of data management is to increase the potential of existing data for company-related objectives. The entire life cycle of the data in the company is taken into account.

The relevance of data governance increased massively with the enactment of the GDPR, which tightened data protection: With the enactment, many companies were faced with the question of where customer data is located in the company in the first place and how it must or can be managed both from the customer perspective with the goal of data protection and from the company perspective with the goal of data security.

Key data governance issues from my perspective are:

- **Focus**: What data does Data & Analytics focus on?
- **Data quality**: How do I ensure data quality for the intended purpose?
- **Data use**: Which data may I use? Which data may I merge? When may the data not be used?
- **Data sharing**: Which data may be shared in which form with which other actors (internal and external to the company)?

The central questions of data governance make it clear that it cannot be viewed in isolation from the corporate and data strategy. Rather, data governance is a logical derivation from the corporate strategy and the data strategy subordinate to it, since, for example, the question of the focus of data analytics can only be answered on the basis of strategic considerations. The dualism of data governance also becomes clear from the central tasks: On the one hand, there is the question of data protection. Data protection. On the other hand, however, as much data as possible should be used by the customer or about the customer.

Ideally, data governance can be used to create a "data map" for the company that makes statements about the following points:

- What are the sources of our data?
- In which systems is the data located?
- "What insights can we gain from this data?"
- Which business processes benefit from this data?

Structuring the data landscape in a company often proves challenging. Attempts to define accountability for data often fail because the data landscape is very complex and dynamic. However, AI can be used for structuring to "navigate" through the data and get an overview. Tied to an example, this can manifest itself, for example, by comparing two tables from different areas. The automatic analysis via AI can, for example, produce the result that the data consider an identical fact, but from different perspectives (cf. Rashedi & Wernicke, 2021).

In my experience, a mixture of centralized and decentralized forms based on the hub-and-spoke principle is the best way to implement data governance. The central unit has the task of defining the guard rails for data governance in the company. The decentralized teams, on the other hand, are responsible for this,

- enable the individual departments of the company in terms of data literacy (through services, training …)
- to communicate the importance of the data of the respective department for the achievement of the corporate goal and the core processes
- demonstrate what the consequences may be if the data governance guidelines are violated
- increase the scope of data-driven decisions in the enterprise.

The tasks make it clear that the employees in the area of data governance represent a bridgehead between IT and the business departments. That is why they need both an understanding of data on the IT side and knowledge of the business of the business departments. Ultimately, therefore, they have to speak two different languages.

In my experience, it has proven to be helpful in establishing the area to enter into dialog with the employees in the departments and to use their "language" in the process. In concrete terms, this means talking to the departments about data and, for example, using short workshops to demonstrate the added value that data analyses can mean for the department.

In my opinion, there are two challenges in connection with data governance. Firstly, determining an appropriate level of effort for data governance and, secondly, ensuring congruence between data governance and corporate and data strategy. With regard to the effort or costs of governance, it should be noted that these can be amortized very quickly. This becomes clear if we imagine that the people in the specialist departments who are responsible for evaluations would only save 1 day per month for searching for data and information.

References

BARC (2019). *BI trend monitor 2019*. Retrieved November 10, 2021, from https://www.visma.no/globalassets/business-intelligence/nyheter/barc_bi_trend_monitor_2019.pdf

Data Governance Institute. (2020). *Governance and decision-making*. The Data Governance Institute. Retrieved November 10, 2021, from https://datagovernance.com/governance-and-decision-making/

Goyal, A. (2021). *The modern data stack in 2021*. Snowplow. Retrieved November 10, 2021, from https://snowplowanalytics.com/blog/2021/05/12/modern-data-stack/

Hürtgen, H., Kerkhoff, S., Lubatschowski, J., & Möller, M. (2020). *Rethinking AI talent strategy as AutoML comes of age | McKinsey*. Retrieved October 2, 2021, from https://www.mckinsey.com/business-functions/mckinsey-analytics/our-insights/rethinking-ai-talent-strategy-as-automated-machine-learning-comes-of-age

Leyritz, L. de. (2021). *How to build your data analytics team*. Retrieved November 10, 2021, from https://towardsdatascience.com/how-to-build-your-data-analytics-team-1276d6729ac4

Luber, S., & Litzel, N. (2020). *What is automated machine learning (AutoML)?* Retrieved October 2, 2021, from https://www.bigdata-insider.de/was-ist-automatisiertes-machine-learning-automl-a-896975/

Momtasian, M. (n.d.). *What is talent management and why is it important?* Expert360. Retrieved October 2, 2021, from https://expert360.com/resources/articles/talent-management-important

Rashedi, J., & Beer, R. (2021). *Romy Beer & Jonas Rashedi - B2B goes digital*. MY DATA IS BETTER THAN YOURS. Retrieved November 10, 2021, from https://mydata.podigee.io/42-b2b-goes-digital

Rashedi, J., & Barth, C. (2021). MY DATA IS BETTER THAN YOURS. Retrieved November 10, 2021, from https://mydata.podigee.io/46-christopher-barth-hugo-boss

Rashedi, J., & Damm, P. (2021). *Sport1 (Pascal D.) - Data, the basis for new publisher products*. MY DATA IS BETTER THAN YOURS. Retrieved November 10, 2021, from https://mydata.podigee.io/48-sport-1

Rashedi, J., & Feng, T. (2021). *Adidas (Tiankai F.) - Why analysts can do much more than Excel and dashboards*. MY DATA IS BETTER THAN YOURS. Retrieved November 10, 2021, from https://mydata.podigee.io/47-tiankai-feng-analysten

Rashedi, J., & Malinov, D. (2021). *Lidl (Dimitri M.) - Data to the people.* MY DATA IS BETTER THAN YOURS. Retrieved November 10, 2021, from https://mydata.podigee.io/40-data-to-the-people

Rashedi, J., & Wernicke, S. (2021). *Sebastian Wernicke & Jonas Rashedi - Data Science on the ground.* MY DATA IS BETTER THAN YOURS. Retrieved November 10, 2021, from https://mydata.podigee.io/38-data-science

Schein, E. H. (2016). *Organizational culture and leadership.* John Wiley & Sons.

Ready, D. A., Hill, L. A., & Thomas, R. J. (2014). Building a game-changing talent strategy. *Harvard Business Review.* Retrieved November 10, 2021, from https://hbr.org/2014/01/building-a-game-changing-talent-strategy

Stolpe, C., & Hebing, M. (2020). *Five first steps to a data-driven organizational culture.* Impact Distillery. Retrieved November 10, 2021, from https://www.impactdistillery.com/de/digitale-transformation/datengetriebene-organisationskultur/

Talocka, G., Skyrius, R., & Nemitko, S. (2018). *Business intelligence, organizational culture, information culture, business intelligence culture.* University of Borås. Retrieved November 10, 2021, from http://informationr.net/ir/23-4/paper806.html

US Department of Defense, D. L. (2020). *DOD data strategy.* Retrieved November 10, 2021, from https://media.defense.gov/2020/Oct/08/2002514180/-1/-1/0/DOD-DATA-STRATEGY.PDF

6

Closing Words

Abstract Data has become an indispensable success factor for every company. However, the path to a data-driven organization is paved with numerous challenges. This book presents a process model for the path to a data-driven company and provides recommendations for the design of all relevant fields of action: Which structures need to be created? Which systems and processes have proven beneficial? How can the quality of the data be ensured and what requirements does the data-driven organization need in the areas of governance and communication? And finally: How can employees be brought along on the journey and what implications does the data-driven organization have for our corporate culture? Jonas Rashedi presents an orientation and action framework for the strategic and operational design of the data-driven organization, detached from current technical solutions. Other experts provide concise solution suggestions and best practices on particularly relevant aspects of selected fields of action.

With this book, I hope to have given you ideas and suggestions that will support you in realizing a data-driven organization. It has long been my concern to convince companies and the responsible persons in these companies of the relevance of the topic "data." Data changes the rules of the game for companies and the competition: procedures and recipes that worked in the past and led to economic success lose their validity in today's world.

Companies are therefore challenged to address the data issue and to consider which corporate or divisional goals can be supported by data utilization. Perhaps the most important aspect on the way to a data-driven organization is the collaboration between the BI value chain and the business departments. Neither player can manage the data issue alone. Rather, there needs to be collaboration, taking into account existing competencies: The business departments understand the business better, Data Science understands the data and can provide both systems and methods to prepare the data and create the conditions for the business departments to do their business better.

In my opinion, the barriers to dealing with the topic of data have fallen in recent years. For example, employees generally have a higher affinity for data and working with data due to both their professional and private points of reference to digitization. In addition, the tools available have become easier to use. This makes data more tangible for users and lowers the barriers to data use.

Despite all the negative effects, in my view the Corona pandemic also had a positive impact on the handling of data and the understanding of the importance of data: On the one hand, many companies had to drive their engagement in digital in order to generate revenue at all (e.g., establishing an online sales channel). On the other hand, the pandemic has made subjective forecasts and expectations based on personal experience and knowledge of the market obsolete in terms of sales and revenue figures. However, data could help in this situation to make statements about the development in the short term.

In my personal opinion, there is no way around the data-driven organization for any company in the medium term. The amount of data available is growing daily—and with it the opportunities for companies to generate competitive advantages through operational and customer data.

To do this, companies need a clear objective and data strategy, as well as implementation within the company, as I have shown. In my opinion, sitting it out won't work and harbors a great danger: the longer you hesitate, the more of a lead the competition will build up. And this is either very difficult or impossible to catch up with. In this respect, the question is not *whether* a company should deal with data, but only *when* and *with what specific objective.*

Currently, many companies are already on the "journey" to becoming a data-driven company, as many managers have understood the necessity of a data-driven organization. However, the understanding that data is not just part of a process, but an independent asset that is important for the company, is central. In the future, it will become clear which companies have this understanding and which do not.